Urban Beekeeping

A Guide to Keeping Bees in the City

By Craig Hughes

The Good Life Press LTD

Urban Beekeeping

A Guide to Keeping Bees in the City

By

Craig Hughes

Published by The Good Life Press Ltd. 2010

ISBN 978 1 90487 1 699
A catalogue record for this book is available from the British Library.

Published by
The Good Life Press Ltd.
The Old Pigsties
Clifton Fields
Lytham Road
Preston
PR4 0XG

www.goodlifepress.co.uk
www.homefarmer.co.uk

Front cover designed by Rachel Gledhill
Printed in the UK by Scotprint

FOREWORD

We have both enjoyed Craig's visits and programmes on the BBC here at Radio Lancashire. We have both learnt so much not just about bees but also about honey and the bees' characteristics; we particularly liked the piece about Italian bees enjoying a good sleep. Craig is incredibly enthusiastic about bees and is not only a real campaigner for bee health and support, but also promotes the benefits of honey for health and supports the Rosemere Cancer Trust in Preston. We can see that this book is full of his enthusiasm, knowledge, care and dedication and reads almost as if he is talking to you.

Ted Robbins, Actor, Television and Radio Presenter, Comedian.

Alison Brown, Journalist, Radio Producer for Ted Robbins.

PREFACE

The history of bee keeping is long and varied. The amount of folklore that surrounds this field is immense and ranges from the practical to the mystical. Bee keeping has been described as the second oldest profession in the world and there has certainly been written more about the second profession than the first. Cultures from the north of Europe and the Russian Steppes to the far south of Argentina have a history of bee keeping, and their myths, folklores, histories and legends are all intertwined with the importance and value of bee keeping to society. Some myths and legends appear to have a strong and deep-rooted approach to nature and value the importance of the seasons, whilst others appear to be fanciful. The old tale that a year with thirteen full moons will bring a wet and cool summer appears to ring true. A fairly common tale was that those infants whose lips were touched by a bee would become great and prolific speakers, storytellers and philosophers - perhaps that is why many famous poets and writers such as Virgil, Sophocles and Plato were all associated with the bee. Bees were often called the "birds of the muses".

Over recent years bee keeping has entered something of a renaissance. As bee numbers decline and imports become expensive, a homespun transitional movement is calling for a greater opportunity to produce their own crops and produce. A wave of creativity has once more embraced parts of the world. As many of us live happily in the towns and cities of the world - with only the occasional glimpse of the countryside from a high speed train - the rise of the urban bee keeper has been most noticeable. There are many city farms that offer courses in sausage or bread-making or give the opportunity for one to work with animals. But this is for the few and it is highly unrealistic for someone to keep a pig or a sheep in their back garden or house. Yet bee keeping gives the individual the opportunity to work with his or her livestock for the full year, on top of a building, in a city centre garden, behind the pub - wherever you choose. It is the only farming activity in a city or town centre that allows you to see the finished product; to start in the spring with your buzzing colony, to end, in the autumn, with drawing off your honey crop. And let me say, your first jar, your first taste and your first sting will always be memorable ones. I still have my very first jar of honey that I produced: I can still remember the first time I ran my finger along a full frame of golden wild flower honey and I can still remember my first sting. It was in fact several stings, all on my bottom lip. As the honey that I had just taken from the hive was "stolen" from the bees, they came in force to track down the culprit, hence the swollen lips. But it still tasted magnificent.

As an ex- lawyer and chef I always had an idealistic approach as to how bee keeping would be. I decided after a near-brush with death that the life of law was no longer for me; in fact I had cancer in three different places. I had kept bees since childhood and always found them fascinating and loved them dearly. When I started to regain my strength post-op, I decided to make my plans. I was going to find my "third way" in life: making a living but not making a fortune; taking from society and putting something other than direct taxation back in. One day I found myself on top of a moor, purple heather all around me. It was early evening in late August and I had parked my truck not far from the hives; they were not bothered by me as I was out of the flight line. I removed my veil on the bee suit, removed my gloves and sat for a couple of minutes to watch my friends buzzing around me. The soldiers came out to see what was what but quickly left me alone. I put some water on to boil and made the obligatory tea, cold milk, no sugar and a single stir. I was on my own with the exception of several million bees. Knowing that 800 to 1000 stings could kill you makes you appreciate and respect the bee. Tea ready, I opened my book, sat with my back to a dry-stone wall and opened my sandwiches. Cheese and pickle. I lived off cheese and pickle at university. I vowed never to entertain it again. Yet here I was, with the warmth of the late summer sun, the glow of a job well done and the satisfaction of knowing that my little chums were busy making honey, eating my cheese and pickle. I had been working those bees all day since first to last light. I had not seen a single soul. You may wonder where that fits in with a book about urban bee keeping. I was asked to get involved with a project in Liverpool and Manchester. The Liverpool project was for an organisation keen to turn around the urban and social decay that it found itself in. The project was no further than one and a half miles from the Liverpool waterfront and its famous Liver Building and Birds. The bee hives were to be sited on a flat roof, on an old pub in a run-down area in need of some care and attention. The owners of the properties, the neighbours and urban activists were superb. They supported the project and once I had spoken and discussed and dismissed any fears or concerns about being stung, swarms and the like, we went ahead with the project. Looking around we could see the trees and flowers, shrubs and plants that would be contributing to our honey. By the end of the summer I found myself sitting on a roof top, drinking tea and eating my sandwiches (yes, cheese and pickle again), looking out over the River Mersey, over to the Wirral, surrounded by the buzzing of several hives and realising that it wasn't the location that had made me realise what a wonderful and rewarding thing I was doing. It was working with the bees; working with nature; and working with keen, like-minded individuals who wanted the best for themselves and their community. There are now bee hives spread out across the north west of England, not only in the countryside, but in the major towns and cities.

For Spike,
for all your help, support, patience
and most of all,
love

CHAPTERS

INTRODUCTION

I have divided this book into what I feel are relevant sections. Relevant because they take in the months of the calendar and for the novice bee keeper we will start logically with what should be your first month. The book will move through what you need to do and when, using tips that I have found to be useful and identifying savings where they can be made. Bee keeping can be an expensive hobby and your initial overheads for starting up can be quite costly depending on how you want to pursue your hobby. You will see I have started with September and not a spring month like April or May. I can guarantee that if I did I would not be giving you time to prepare and this book is all about preparation. It's what we call the six Ps and a description of these can be found on my website, www.crossmoorhoney.com.

You are about to embark on an honourable and ancient profession. You will receive lots of advice about where to place your bees, what queens to use, what type of hive to use and so on. I can only guide you. What I can say is use what you feel happiest with. An example of this came to me when a friend asked what was the best type of hive to use? I asked him where it was to be positioned. He wanted something old and English which was to be sited in a country garden. In this case the best hive, aesthetically speaking, was a W.B.C. These are very old, very heavy and cumbersome. They also are renowned for draughts.

Personally, I like to use a polystyrene Langstroth hive. Now I don't want to frighten you and become all technical. Langstroth is an American vicar of the mid-1800s who designed a wooden hive. Nowadays you can get them in polystyrene which is a strong, robust and very easy hive to use. But I have found some people do not like to use anything else less than wood and a design that bee keeping clubs are use to. If you are placing a hive on top of a building, my advice is to use a polystyrene one and anchor it down. It will be so much easier to transport if need be because there is always the question of weight. Of course a wooden hive on a roof, for instance, will appear more secure than, say, a polystyrene one. However, open to the elements, a polystyrene one will last much longer, is easier to move, has less chance of getting disease in it, and the pieces are cheap to replace. If the worst happens and a polystyrene hive blows away it will break up a lot easier than a wooden one and cause less damage - especially if it's heading for a passer-by.

Chapter One

THE BEEKEEPING YEAR

* * * * * * * * *

SEPTEMBER

If you are just finishing this book you will be smearing the first of your golden harvest over soft warm bread, stirring it into deep, Greek, fruit yoghurt or just simply licking it off your finger tips. Either way you will be enjoying the fruits of you and your buzzy little friends' hard work. If you have not got to this stage you will have the fun of years of bee keeping ahead of you.

First of all I shall cover some basic advice which you will find under the following headings:

> **Read this book**
> **Get stung**
> **Look at the deeds of your property**
> **Identify your site for your bees**
> **Buy your equipment and familiarise yourself with it**
> Contact your local bee keeping association about buying some bees, or a breeder for a nucleus
> **Watch the weather**
> Have a plan B

Read this book

By reading this book you will be guided page by page through the pitfalls and benefits of keeping bees. Keeping bees is not just a benefit to society, the world or the environment, but also a direct benefit to you by eating that sweet, sticky nectar. If you have a garden, fruit trees or soft fruits, you will find that bees, once they have started pollinating the crop, will increase your productivity by 18% to 36%. That means the possibility of an extra 36% raspberries or plums or even courgettes. (You will wonder what on earth to do with the abundant crop of the latter!) I have used my experience as a commercial bee keeper to open up the opportunities to those of you starting out. With colour photographs, hints and tips you should be able to enjoy years of problem-free bee keeping.

Get stung

Many people believe that they will suffer from anaphylactic shock (A.S.) or form a severe reaction to a sting. One in 100,000 suffers from A.S. But many are frightened of the possible consequences if they are stung. They are catastrophising the situation and are allowing their concerns to colour a beautiful pastime. The best way to ensure you are suitable for bee keeping is to get yourself stung by a honey bee. Do not try it with a wasp or a bumble bee as the effect will be totally different. Visit your local bee

Urban Beekeeping - A Guide to Keeping Bees in the City

keeper or bee keeping association and ask to be stung. Explain why, and they will be prepared if you start to struggle. In over thirty years of bee keeping both commercially and as a hobbyist I have rarely come across someone who reacted badly.

Late one evening I was moving some bees from a hedgerow to a field of courgettes. I dropped the hive, heard a lot of buzzing and in the dark I could not see where the hive had fallen, so I left it. I retreated to the truck which was now buzzing loudly from the inside where other bees had sought refuge! I knew it was time to leave and tried not to panic but as the bees began to sting I knew that a sting on the throat, eye or lips had to be treated with concern. I managed to drive around a mile away and get out of the truck. I lost count of the stings after I passed the 250 mark. My wife kindly removed the other

sixty from the back of my neck, scalp and back. I returned to work, a little stiff but well enough to continue. On another occasion a young woman was helping me with the bees. I had explained about the bees and that a certain hive was particularly aggressive. I expected her to ensure her suit and veil were closed. I should have checked because when she panicked and dropped the hive, a column of bees shot up her suit and into her open face - she had not in fact secured her suit and hood properly. She rolled around the floor as the bees attacked her. I was able to pour water over her face and get her into the truck and drove a distance away. Killing the bees with my hand I was able to remove as many as possible from her face but by this time she was unconscious. I raced off to a local hospital some 20 miles away. After an injection of antihistamine and a lie down she was fine and was back to work the next day.

She had had around 30 stings to her face. She said that she did not want to fully close the veil as it damaged her hair! A lesson learnt.

Look at the deeds of your property

If you live in an old property, have a look at your deeds. If your property says you are not allowed to keep livestock, and in particular bees, then unfortunately that's the end of bee keeping at your property. If yours is a rented property simply ask your landlord to clarify the position. The reason why I mention this is that it will be an expensive adventure if you get so far but then cannot keep bees purely because of an administrative problem. Most old houses do not mention the keeping of bees nor any other animals, whilst some do note that there is permission for chickens, bees and even pigs in some instances. If there is no mention that you cannot keep livestock, then it is generally accepted that you are allowed to. Most modern buildings, however, certainly those from the mid 1950s onwards, will have a note on the deeds forbidding such activity. It pays to check first. Always keep in the forefront of your mind that bees, like any livestock or living creature, need several items to survive, let alone flourish: they need shelter, food, warmth, water and light.

Identify your site for your bees

When you are considering where to place your hive look at what is around you at the time. In September you should be able to

get a good idea as to what is in your locality and remember, bees will fly up to three miles away searching for pollen and nectar, but it's always better to place them where they are near their source of forage. By having a quick glance you should be able to see what is available, where the best shelter is and where the best access is. Look around, not necessarily at what you are growing, but what your neighbours have in their gardens or what is growing in public places, and you will be able to decide if there is enough forage for the bees. As I look out of my window now I can see many broad leaf trees - sycamore, ash, elder and oak. There is also a hedgerow, clovers in the grassland and pansies and small spring flowers that the council have planted. In addition there are foxgloves, hollyhocks, sunflowers and lupins in my neighbour's garden. Commercially a bee keeper will look for vast areas of plantation - rape, lupins, beans, sunflowers and heather - but from an urban bee keeper's perspective you will need to concentrate on what is available as you won't be moving your bees around. Remember the cities and towns are 2-3% warmer than out in the countryside, are more sheltered and therefore have a longer season. This means that in mid-August when I am moving bees away from the lowland crops onto the moorland heather, you won't need to and won't miss out on local autumn flowers which will still be flourishing in the city gardens, especially those like ivy which gives a beautiful honey. (There is a section later in this book about gardening for bees.)

If so far you have ticked all the boxes, look for a place to site the hive. You want somewhere sheltered from the prevailing wind, which is usually westerly, where your hive can face south and benefit from the sun from sunrise to sunset. You may wish to site your hive on a roof top, a bay window or in the garden or allotment. Wherever you choose, you will need to place the hive bottom on something that raises it off the floor. This is to allow ventilation to run underneath the hive and allow fresh air to flow. This minimises disease, prevents grasses growing through the base of the hive, prevents other insects from gaining a foothold and keeps the damp out of the hive. Bees really dislike the damp and wind and if you can eliminate these two elements you are part of the way to keeping your bees alive over the winter. When you are looking to raise your bees off the ground don't go to any great expense. More on this later on in the book.

Buy your equipment and familiarise yourself with it

The right tools will always help. When you buy your equipment, buy it sooner rather than later, as you will find that around springtime certain items may be out of stock. Essentially, (and to keep your costs down) you will need the following:

- a hive tool
- a bee brush
- a comb
- a smoker
- a hive
- a bee suit
- a travelling strap
- bees

Hive tool - claw

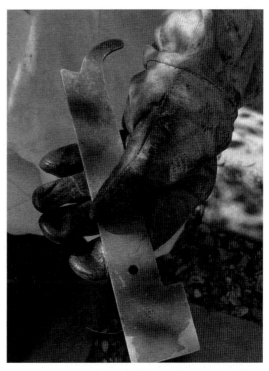

Hive tool - J type

I have suggested a hive tool. You will find that if you buy just one you will invariably lose it straight away. I would buy two to be safe. In fact, at the start of every season I buy eight just to be on the safe side.

A bee brush is used to remove bees gently from frames or other areas without damaging the bee or upsetting them too much.

Your comb is used to lift the drone brood to check for varroa. You will really only need one of these as they are not used too much over the season.

When buying a smoker, get a robust one - one you can drop. Also, make sure it has a good set of bellows that is easy to clean and has a guard around it where you can place your hive tool when you have finished with it - that way you will always know where the hive tool is.

As for smoking materials you will have a plethora to choose from. One rule of thumb for me is always to use what you have to hand. You can buy really expensive inserts for your smoker and everyone will tell you they will do a really good job. A word of warning, though. If my wife drinks red wine, her legs jump around uncontrollably when she goes to sleep. Red wine doesn't affect everyone like this, just my poor wife. Bees are no different. If you smoke them with one product you might find it works well, another not so. But bees are woodland creatures and will react directly to their environment, so choose a product that they are likely to encounter. Essentially, when you are smoking the hive, the bees believe that the forest is on fire and their safety, the Queen's especially, is under threat and they must move. So ideally, choose something that you will find in the forest.

Hive tool - comb

I like to use a combination of items. I use peat and very dry tree bark with occasionally a dry pine cone thrown in. I find this combination works well: the cool smoke from the pine cones calms the bees and the peat burns nice and slowly. Bee keepers are usually very conscious of the environment and of living as sustainably as possible. It would therefore be silly to drive thirty miles to a pine forest to collect cones for burning. Often you have to use what is at hand. For example, if you get a delivery in a corrugated cardboard box, discard all the plastic and non-organic packing materials so that you are left with the cardboard. Cut it into strips that will fit neatly into your smoker, allowing enough air to circulate around and light it. This works well but I have found it can have a tendency at times to annoy the bees. Tree bark, wood shavings, even coconut husks, work a treat.

Two types of smokers

(But unless you are reading this book and keeping bees in the Caribbean you might find husks difficult to get hold of.) You can also use jute or hessian sacking which are both fairly easy to obtain (I get mine from my local pet store). Sacking is effective on the bees and will calm them down quickly, but you must make sure that the sacking has not been treated with a fire retardant which can be carcinogenic if burnt and breathed in. You can use grass if it is dry; on its own it burns very quickly, but mixed with coarse wood shavings it can give a good, dark smoke. If you have a garden centre close by that chips its own wood, or cuts wood for burning in fires and stoves, more often than not you will find there is a surplus of bark that they can't use and are quite happy for you to take away. Equally, if you live close to a chicken farm ask them for any damaged egg boxes and/or chicken feathers, but remember to mix the two as the feathers alone will burn rapidly and leave you wanting - and they also have a very distinct aroma when burnt.

I mentioned to a fellow bee keeper the other month that I was writing this book and I asked him his thoughts on smoking. He told me that he lived close to a dairy farm and collected dry cow pats which, when mixed with dry horse manure, make a very effective smoking material. He told me that he had got the idea from witnessing Arabs in Algeria doing the same with camel dung. I have also been told that tobacco waste works well as does hemp in its various forms. Apparently, but perhaps not surprisingly, it sedates the bees.

What you don't want to be doing all the time is stopping to relight your smoker over and over again. From a practical point of view this can be a nuisance, to say the least. Imagine a scenario. You have thick gloves

Bee keeper holding
super frame (shallow)

on so you have little or no feeling in your fingers and you are trying, unsuccessfully, to roll the barrel of the lighter, so you take the gloves off and you get stung. Or you are using surgical gloves, and the flame keeps going out and your gloves start to rip. It is extremely frustrating. It is far better to work with something that is going to keep on burning even in a wind.

In summary, then, a good smoker fuel should have the clear characteristics of being non-toxic, cheap, easy to obtain, easy to light and stay so, and importantly, present a lot of smoke. Once you have decided on your material the first secret is always to make sure it is dry; the second secret is to have a good supply of material available. And the third secret of smoking success is to present a jar of honey to the person who has supplied the material at the end of each season.

The hive is obviously a vital piece of equipment. You can choose from a W.B.C., a National, Smith, Commercial or Langstroth. (To find out about the different hives that are currently available have a look on the internet.) All have their pros and cons. A major factor if you decide to have a wooden hive must be the cost. A wooden W.B.C., for example, (see page 82) will be several hundred pounds, a polystyrene hive will be in the region of £75. And remember that does not include the bees. I have used the W.B.C., National, and Langstroth and have always had good 'honey' results with all of them. When you are talking about all year round performance, however, I would favour the Langstroth and especially the polystyrene type. There are two main reasons which

make me err on the side of polystyrene. Firstly, I have always found that if you can keep bees away from anything organic, like wood, you reduce the risk of disease a little. Secondly, I find that if you are moving, inspecting or replacing parts of a hive, it's a lot easier when you are using a material as light as polystyrene. I do keep polystyrene hives on roofs in very exposed places and, as long as they are well tethered, they will be safe enough.

Whatever hive you choose to buy, give thought to the type of frames you are going to use. If you are in an area where there is a good flow of honey, you might be able to manage with plastic frames. These are good because you don't have to worry about

Langstroth hive made from polystyrene

ROOF ⇨

SUPER ⇨

BROOD CHAMBER ⇨

FLOOR ⇨

framing and re-framing over the years. You can just spray a fine layer of hot beeswax over the frame on both sides and allow it to cool. But if you live in an area where the honey flow is slow or sporadic, you are better with the traditional wooden frames with the wax pressed over wire. Again a rule of thumb for dealing with the plastic/wooden dichotomy is that if you are on the west of the Pennines you are better with wooden and if you are to the east then plastic is a better option. A drier climate is usually better suited to plastic frames. Time for tea – a bee keeper's best friend.

A bee suit must be something you are comfortable in and with. Always err on the side of room - if you think you will need a medium size, opt for large. The reason behind this is that many people have nice calm bees that work well and only occasionally sting. If you are fortunate to be in this position then you may well get away with wearing just a t-shirt and a bee suit. Unfortunately, I have bees that will sting you for fun. They tend to be aggressive and sometimes downright unpleasant. But they are hardy, hard-working, survive a northerly climate and I love them. But because they can be belligerent I wear several layers - a t-shirt with long sleeves, a pair of overalls, thick socks and a neck scarf - and then my bee suit over all of this - and Wellington boots. It may appear excessive, but if you want to keep them out, suit up. Even if your bees are not as feisty as mine it is still worth wearing an extra layer. What if it rains and you have to travel some distance to get home? At least you can remove your outer sticky, wet bee suit. You may have noticed I've not mentioned bee gloves. You can buy a really nice pair for around £35-£40. They will be leather gauntlets, and very thick and hard wearing. One problem with the leather glove, however, is that they are invariably yellow calf skin. The dye will transfer onto your hands and you will look like a chronic smoker with terrible nicotine stains. This is a minor nuisance compared to the lack of ability to feel what you are doing. Some jobs require sensitivity and dexterity. If you are marking a queen, for example, you don't want to crush and kill her. My advice is to give the leather gloves a miss and buy some latex surgical gloves. I always wear two pairs. This prevents any stings from getting through to the skin and really allows you the luxury of touch. Another major advantage over leather gloves is that you are able to change and dispose of surgical gloves regularly which will reduce the amount of potential disease being passed between hives.

A travel strap is another piece of equipment that you might like to invest in. There has always been a dispute between whether you should use a travel strap to hold the hive together or whether you should just use a brick on the top to weigh it down. Both have their merits, but I find if you use a strap, it keeps the hive all in one piece should it be knocked or blown over. The down side of this is that should someone decide to steal your hive then a pre-strapped one is easy prey. A hive with a brick is much less easy to steal but is prone to movement in the wind.

Contact your local bee keeping association about buying some bees, or a breeder for a nucleus.

I will deal with types of bees later on in the book but a word about queens. If you are looking to buy a queen, try to find out about

the breeder. How long have they been breeding bees or queens. Buying a queen is very much like buying a pedigree puppy. Buying a dog is a big expense and you would do some investigating before you commit to buying it. Has the dog come from a loving home; is it a puppy farm; are the parents in good health - you can see where I am coming from. Well it works the same with a queen, but in reverse. Your outlay for her may be anything from £8 to about £35. It's a small sum if she dies, but if she is highly productive and produces aggressive offspring then the cost of replacing an entire colony is prohibitive. (You may be forced to get rid of an aggressive colony if your neighbours complain.) But perhaps the most important question is, has there been any disease in the area that I am buying my queen from? In particular, it is wise to try to identify whether there has been any reported AFB (American Foul Brood) or EFB (European Foul Brood). You can check this with the National Bee Unit or with your local bee inspector.

You can also buy a nucleus, or a small half-hive. You can do this by contacting a breeder or through a bee keeping club.

Weather watching

You will need to buy a book on the weather and certainly to watch the weather reports at the end of the news every night as this information may well dictate whether or not you can work your bees. Even better than watching the weather is being able to predict the changes. By looking at clouds and wind patterns you will be able to identify what is about to happen and when. I'm not going to go into detail about the jet stream or the difference between cirro-strata or altocumulus as that would become a totally different book all together. But if you work with the weather, you will succeed where others have failed. My father was for many years in North Africa. He worked in the water supply industry which was of real benefit when you are based in the Western desert for months at a time. He never wore a wrist watch and would take what others thought was inappropriate clothing out with him on his journeys. He would often find that his 'inappropriate' umbrella and overcoat came in extremely useful when the wind picked up or a sudden downpour occurred. His skill of telling the time by looking at the sun and identifying possible bad weather have all been passed on to me. Of all the many pieces of information my father passed on to me, 'weather watching' has been the most useful.

Tips!

- *Never forget your matches!*

- *Never tell the wife!*

- *Never go out without the Plan B!*

Plan B

Take nothing for granted is a rule of thumb when dealing with livestock and always have a plan B.

I can not say to the letter what this should be other than you should always work on the basis that something other than what you have planned is bound to happen unless you have all your equipment or a contingency plan in place. Let me explain further by giving an example.

I had 75 hives to transport up the moors in Lancashire. I arrived, removed the ropes and packaging for so many hives, carefully placed them on the quad bike and travelled steadily and slowly a mile away. Gingerly feeling my way and cutting a path through the lowland heather I eventually placed my first three hives 20 minutes later. To get 75 settled in one day was a long job and a long day. To finish off I had to have a quick look in my hives just to see the state of play.

Knowing that the bees would be angry with me for jiggling them around I decided that the best course of action would be a good puff of smoke to keep my little chums content and quiet. Before I did this I would make myself a nice strong cup of tea, have a sandwich and revel in the delights of a job well done. I filled my billy can from a stream and took the paraphernalia of survival out of my "essentials" case only to find that I was missing the very thing that sets us apart from the animals – fire. There was no lighter and no matches, not a spark for either the smoker or my tea. I was on top of the moors, looking over the Irish Sea and I could see the clouds of a storm steadily working its way across the water and directly towards me. I could not leave the bees bunged up; they had to have fresh air and freedom, but they would also be very cheesed off with me. I couldn't leave them enclosed, but nor could I open the hives to see if they were fine. All because I didn't have any means of lighting the smoker. I packed everything up and headed off the hillside to be greeted shortly afterwards by a heavy downpour. By the time I reached the bottom I was wringing wet. I had to get the quad up on to the trailer. As I moved up the rails the quad slipped, I slipped, and the quad fell of the rails and on top of me, pinning my

leg underneath. Had I been able to smoke the hives and make some tea I would have stayed up on the moors and sheltered. By this time the quad was leaking petrol and I was somewhat worried. I was able to get the quad off me and eventually onto the trailer. I drove home and like all good husbands didn't tell my wife what had happened. Because of my accident - and no plan B - I lost several days work. Just to set your mind at rest though, I did manage to return to the bees to unbung them and all was well.

OCTOBER

For those of you with bees you should by now have placed your bees somewhere safe and dry and away from the prevailing wind. And you must have fed them. If you are taking honey or store feed from them, they will need something to eat in place of what you have taken.

It is better to give your bees a good feed around the end of September, after they have foraged for any pollen and nectar that may still be available. You can use a home-made mix of sugar and water. To prepare this you must boil 2 parts sugar to 1 part water together. Let it boil away until it reduces and forms a sticky syrup. Allow it to cool and pour into your feeder. Simple. Well, not quite as simple as it appears. Unless you get the syrup to just the right level of reduction and all the water is removed you will find yourself with a solid mess in the feeder in mid winter. Your bees will have a frozen block above them from which they can't feed. It will also lower the temperature of hive and will probably cause the colony to die.

Alternatively you can buy in some feed. There are two types: fondants and fluids. Fondants are a block food which is slipped into a sleeve that looks like a normal frame and which sits next to the brood. The bees move laterally to find the sugary food and 'Bob's your uncle'. Or if you are reading this book in Italy, Roberto et to a zio. Fondant is not as sticky or messy as other feeding mediums. Over the years I have found that bees like to move vertically: they will always look up and work upwards. Because of this I like to use a reservoir feeder. This is a feeder that sits on top of the hive, in place of the crown board, and holds the liquid feed. This allows the bees to make their way up to the feeder and sip the liquid without getting themselves covered in feed. You will find that most commercial feeds have approved additives and medicines in them. The only practical way of treating bees is through their feed - you can hardly give each bee an individual dose of medicine as you would a cow or sheep.

So feeding, in short, is a block of fondant, or a couple of gallons of liquid feed into a reservoir, lid on, strap on, and leave over the winter. In the spring give your bees another good feed to kick start the colony and get Her Majesty laying eggs.

A word about the siting of the hive. The thing about bee keeping is that it's one of those hobbies that people are curious about. Ninety percent of the population have probably never seen a hive, although many will have been stung and tasted honey. So if you are placing your hive in the city or town, the rule of thumb is to either hide it away so that it's completely out of site, or put it in full view. Don't do it by halves as you will encourage the curious to venture too close and they will invariably get stung.

Personally, I always like to place the hives out of sight. I try to keep them away from footpaths or tracks (this is much easier if you are siting a hive on a roof!). However, if you do leave your hive in a public place, close to a footpath or some other right of way, and someone does decide to have a look inside, as long as there is no signage telling them to stay away, (Danger Bees or Stay Away) then you can not be held responsible if they are stung. The fault will be theirs and they will have to prove in court that they are in fact your bees. I know that appears perverse, but you could find yourself in deep trouble if you decide to bring your hive to the public's attention.

If you site your hive under a hedge or tree, make sure that any branches or boughs are not knocking on the hive in the wind. This will cause the bees to keep coming out to see what is causing the vibration. In the winter this will prove fatal as they will undoubtedly die as the cold hits them.

NOVEMBER/ DECEMBER

These are months of more preparation rather than working with the bees. There is much that can be done. You should use this time to acquaint yourself with the joy of disease control and hive preparation. In these cold winter months you will be able to get a lot of those jobs done that you really need to do but have put off during the earlier part of the season. I always like to prepare any hives I am using for next year - a clean hive is a necessity. Insects like spiders are no problem in a hive, but others, like wax moth and varroa, are. Little mites, and especially wax moth larvae, tuck themselves away and can live quite comfortably totally undetected. Unfortunately a strict regime of bug and disease control also means that good guys, or at least guys that do no harm to the bees, will also be destroyed along with the bad guys. Nevertheless, this is the price that must be paid for a healthy hive.

One of the easiest ways of ensuring that your wooden hives are disease and insect free is by using intense heat. This can be done using a small burner, like the type you would use to remove paint from a window frame or start a barbecue with. You should clean all parts of the hive. Ignite the flame and run it across the wood until it scorches, making sure you go into the cracks and creases of the wood. Then give the hive part a good shake and you will find most of the debris will fall off: just give it a final clean up with a cloth and you are ready to go.

If you are using a polystyrene hive, clearly using a naked flame will be a disaster. A good alternative is to dip your hive parts into a solution of caustic soda (sodium hydroxide) for five minutes - but be very careful and use rubber or PVC gloves and follow the manufacturer's safety instructions on the bottle to the letter. Although this treatment will pit the surface slightly, give it a generous coat of outdoor paint and it will be as good as

new. Alternatively you can dip the hive parts into a sterilising solution such as Milton for at least fifteen minutes.

Whilst you have a little more time on your hands, why not get some of your other jobs done. Making and waxing frames is a good task to be getting on with at this time of year since any wax moth larvae will be inactive and won't be attacking your wax foundation sheets. Make your frames up by simply inserting the sections into each other. Start with the top bar and place the two side bars at the head of the top bar, insert them into the slots and finish with a couple of tacks or panel pins to secure them. You are always better fixing the frame with pins because when you pull one fully laden with honey from the hive it can be deceptively heavy. If your frame is not secure then you are likely to witness the unfortunate experience of watching all your lovely golden honey fall from the frame and land on the ground in a gloopy mess.

Once complete your frame will need some wire to which to secure the wax and, ultimately, give structure to the honey. At each end of the side bar you will see a number of holes. These should be filled with little brass eyelets which can be obtained from most bee equipment suppliers. Now comes the wiring stage. Run a length of wire (I like to use 0.4mm wire) through the brass eyelets until you can secure one end to the outside of the side bar with a panel pin. Do this and then feed the wire back through. Then cut the wire, pull it tight and secure it so that once it has been secured you can pluck the wire and hear a twanging sound come from it. (You might like to have a generous amount of fabric sticking plasters available because trying to get the tension right on the wire has a tendency to cut your fingers - and always at the joint.) Once you have wired your frame you are at the waxing stage. Place a sheet of wax foundation on top of the wires. Now, there are several ways you can get the wax onto the wire. A good strong hot blast of air from a hair dryer will work but is not the most effective or reliable way. I find the easiest and most successful method is this. Buy a cheap battery charger for a car (and a handful of 13 amp fuses) and plug in. Take your frame and place the wax on top of the wires. Now attach either the positive or negative terminal head on to one of the exposed panel pin heads. Then carefully place the other terminal wire on the other panel pin head, making a circuit. The circuit will then heat up and 'hey presto' you have a hot circuit. It will only take around three seconds for the wire to heat up and the wax to settle on to it - leave it too long and the wires will melt through the wax completely and it will be of no use. Remove the circuit wires and place them to one side, avoiding contact with anything else that is metal. Put your completed frame into a clean brood or super box to avoid it becoming damaged. The reason why I said to buy a handful of 13 amp fuses is that the car battery charger is not designed to be used for waxing frames so it has a tendency to short out every now and again. It's always better to have a few fuses to hand when you're doing this job otherwise you will find yourself visiting the local DIY store more times than you would care to in one week.

As well as carrying out jobs inside you should also still be checking on your bees. When you do, make sure that you place a mouse guard on the front of your hive. This does exactly what it says it will. It protects the colony from mice and other small rodents breaking in and robbing wax and honey. Mice guards are fine for wooden hives, but are a waste of time if you use polystyrene hives as the rodents will just gnaw their way around the guard and into the hive. There is the belief that as the rodent chews its way into your polystyrene hive it will make itself ill on the polystyrene and not go any further; this may work in theory but I'm not so sure about in practice.

Another unlikely pest is the woodpecker. You may think it's implausible to suggest they will be found in an urban situation but a city park is as likely to support woodpeckers as a small country copse. To avoid damage by these birds, especially in spring, place some chicken wire around the top of the hive - it won't hurt the birds and it will keep your hive and precious bees protected.

JANUARY

Ah yes, the season of good intentions. I once joined a gym - never went. I also joined a fell walker's group, out on the hills and mountains, me and the bracing elements - never went, too cold and miserable. But now is a good time to start with your bee preparation - and it's a lot less energetic than my other aborted plans. Start by ordering your hive and your other equipment and by making enquiries about obtaining some bees in the spring.

Have a walk amongst your hives and check for any damage by pests - if you didn't fit mouse guards last month you may find some damage. Unfortunately you may find damage from the worst of all possible pests - people. Personally I have known members of the public to damage hives, playing a form of 'dare' with them. Once they have been badly stung they come and complain to you and try to make out it's your fault. Youths have been known to turn hives over in the winter, throw bricks at them and generally be an all-round annoyance.

Whilst on your walk about you should also lift your hives from time to time without disturbing them too much and assess whether they have enough feed. The ease with which you can top up their provisions depends on the type of feed you have given them already. With a reservoir feeder you can merely remove the brick weight or strap, crack open the roof, look inside and replenish as necessary. If you are using a fondant, however, then you may need to open the hive up if you feel there isn't enough for them to eat. This is a task that I would not tackle this early in the year as you are losing your heat from the hive and causing the bees to stir and fly out. In January you may find the weather is warm (for winter) and there is a slight breeze from the Sahara, pushing temperatures to the 10-12 degree mark. At the same time you might easily find that the temperature is at freezing and doesn't lift for weeks.

Remember, however, that EAT=HEAT. If they don't have enough food, it is too far away for them to access or stock is running low, the colony will start to suffer and may die, so you may well have to open the hive, even against your better judgement.

If your hive feels full of feed or bees but you want to make sure, then rather than taking the top off to sneak a look, simply have a listen. Try to pick up an old stethoscope (or even use a jam jar) and listen at the side of the hive. Give the hive a rap with your knuckles and wait and listen. You should be able to hear the sound of fed up bees being disturbed. If the area you live in is prone to snow, make sure that the landing board isn't obstructed. Also, clear way anything that should not be there, including dead wasps, bees and bird muck.

FEBRUARY

This can be one of those months when you can either get on with a great deal of work, or the weather will defeat you. If you have not checked the level of food stores in January, then now is a particularly good time to do it. It is a fact that bees are more likely to die in late winter and early spring than at any other time of year. It may be considered odd that bees would die in the later half of the cold season but you may find that even as late as March bees can starve due to the lack of availability or isolation of food. Imagine that the 'ball' or cluster of bees at the start of winter will be in the middle of the frames, near the front and low to the bottom. As the temperature changes and grows colder the 'ball' will move upwards and backwards away from the entrance to escape the cold. Remember it needs to be 35°C inside the hive for the rearing of brood and when that temperature is not being produced naturally via the sun, it needs to be produced by the bees eating food and generating heat themselves.

You may find that in milder weather, as the external temperature increases, the internal temperature of the hive will increase too and the behaviour of the bees changes. The warmth will encourage the bees to break from the 'ball' and move freely around the hive, preparing for foraging and repopulating the hive. They will necessarily use more energy and will consume more and more of the store food. Some bees may even be tempted outside. If they don't die outside the hive through not being able to sustain the body temperature they had inside, they will jeopardise the colony on their return by eating more store food because there is not enough natural food to be obtained outside the hive at this time of year.

You may have followed all the procedures for supporting your bees only to find they have starved. If they have, don't beat yourself up about it. I always expect to lose about 10% of my bees over the winter. Some are weak, some are old, some are just not able to survive and it boils down to natural selection. When you lose 40% of your hives, however, then it becomes a worry. This is why it's important to keep a check on feed over the winter. Cold weather is not too much of a worry for bee keepers in a temperate climate like the United Kingdom, although perhaps the winter of 2009/10 will prove me wrong. Even the minus 10°C that we have experienced this year is nothing compared to the winters of Canada, America and the mountains of the Alps. All these areas produce many tonnes of honey a year, export queens and have colonies that survive a lot more than a British winter can throw at them, so there is hope for us.

I want to stop here and just remind you why you are doing this activity. I feel I have harped on about the negatives of bee keeping, going out in all weather to see if your hives are upright and the colony is safe, and so on. It all seems doom and gloom, yet once you have extracted your first golden honey and tasted it on your finger, you will never forget it. If you were keeping pigs or sheep you would know that the taste of perfection on your plate would be worth the muddy, frosty mornings for a first class product. And at least as a bee keeper you don't have any muddy frosty mornings to deal with!

MARCH

Let's work on the basis that spring has sprung, the daffodils are out and, with the sight of the first bumble bee, thoughts turn towards bee keeping, honey and the first sting of the year. The air is warm and you contemplate the forthcoming year. The first thing to do is to check the colonies to see their status. What are they doing, what should they be up to and what if they are dead?

Sometimes the bee keeper will feel that his bees have made it through the winter only to find tragedy. A dead hive. As a bee keeper you have to be your own vet. By the time you have identified a problem it's usually too late. So it's important to keep an eye to any potential problems. What you have to do is learn by your mistakes. What am I looking for if I have a dead colony?

1.   Has the colony had enough food? Has the colony's food been robbed by others? Is there any chewed comb? Has there been a rodent in eating the stores and wax? (If you have checked the food and guarded against mice as I suggested above, none of these should be the cause.) Have the bees been kept

from their food for some reason? This is sometimes the strangest and most upsetting of all phenomena - there is food present but the bees have not eaten it.

2.   If there are any dead adults on the floor, then look for disease. Contact your bee inspector for assistance.

3.   Are there signs of yellow spotting? Have the bees had dysentery? Have they had unsuitable food or been kept in the hive because of a blocked entrance and therefore haven't had a good supply of air?

4.   Is the colony queenless? Is there missing brood?

At this time of the year I have occasionally found spotting on a white wall or on the laundry, much to my wife's annoyance. This is likely to be nosema, also known as bee dysentery. It can be a mild infection that clears itself spontaneously once the bees have managed to fly. At other times it can run riot through the colony and reach that critical point where the bees will not recover. Bees can come into contact with nosema when the queen's rate of laying increases and cells are required for the larvae. The bees will clean out old cells in preparation for the laying queen and it is at this point that they may come into contact with the nosema. An infection, even on an apparently small scale, can set you back. Remember that during the season a worker bee has an average lifespan of only 6 weeks. If the worker bees are infected then they cannot attend to or feed the young and the next generation of workers are doomed. What appears to be a buoyant hive with a good body of bees can

be very quickly decimated and ultimately dwindle to nothing. A nosema outbreak can happen to anyone - even after all these years it still happens to me. The good news is that you can treat your bees with a fumidil B substance through their feed, but at this time of year you will have to be quick and warm it slightly.

Let's now look on the positive side and assume all your bees are happy smiling little creatures, waiting to be off foraging and working hard. When the days lengthen and start to warm up, you should see the bees flying freely - it really is a fantastic feeling when you stop and look at a steady stream of bees in and out of the hive. You may well think that, as wild creatures, bees can just be left to their own devices. This is true to a certain extent, but having a hive is really more about managing the bees and providing them with sustenance if they

cannot find it for themselves. This is certainly the case during the winter and spring when we provide them with food, having taken their honey for ourselves. At this time of year it is also about them finding water, or you providing it for them. (See June section for more on water supply.) Your bees will need water to dilute whatever winter food stores are left available. They will also be looking for pollen. Having a good dose of pollen is a bit like humans having a tonic. And if the bees don't have to fly too far to get it, so much the better. So if you have a garden, plant up flowers that will produce a good amount of pollen in spring (see the chapter on Gardening for Bees). If you are close to the countryside or you are not too far from an oil seed rape field, you will see your charges make a bee-line for it as soon as it comes into flower. They love it and will always prefer it over other flowering plants at this time of year. I find that although the bees do enjoy working the rape, it can make them a little difficult, if not aggressive. Rape doesn't make the world's finest honey but it does make a lot of it and bees will build up very quickly on it. So if you can get your bees to it, it's worth the effort.

Before the real fun of the bee keeping season starts, however, there are a number of jobs you can be getting on with.

> 1.   Remove the mouse guard and allow the bees to move more freely.

> 2.   Assess for varroa mite by removing the floor and counting the mites. Treat the brood for varroa at this stage, if necessary. (See section on disease)

> 3.   Replace the floor of the hive. The old one should be cleaned. (See November/December section)

> 4.   Mark and clip the queens' wings. (See section on bees)

You may find that if you have two or more hives, one is considerably weaker than the other/s. If this is the case you may want to think about moving some brood and bees from the strongest hive to the weakest. Before you do this, however, CHECK FOR DISEASE FIRST. There may well be a reason why one of the colonies is weaker. (See Disease section)

What you should do is initially ensure there are enough bees to physically cover the brood when moving the frames; this is an important factor as any exposed larvae will become chilled. If the eggs or brood are sealed it's less of a concern. Next find the queen from the donor colony and place her in a queen cage with a couple of workers. Now decide on a suitable donor frame - one

where the majority of brood is sealed is ideal - and give the frame a light shake over the brood nest, nothing too violent though, and this will dislodge any of the older bees.

At this stage just remember bee keeping can be a fairly complicated business. Don't worry, though, once you have learned how to do this you won't forget it - much like swimming. Once you have selected your brood to be donated, return the queen and close the colony up. By this I mean sealing the entrance to the hive so that no bees can get out. You can do this with some straw or I like to use a piece of foam from an old cushion. Cut a strip of foam to fit the landing board aperture (about one inch square by ten inches long) and push it in the opening. You may need to use your hive tool to do this. Some bee keepers like to dust the bees with icing sugar or flour before introducing the new brood frame, but you need a still, dry day for this to work well. (Dusting the bees will keep them occupied and help prevent any fighting between the receiving and donor bees.) My main concern with using this method is that sugar has a tendency to attract other bees and flour can land too heavily on the bees. If it does decide to rain whilst you are doing this, there is the problem of the water and flour turning to glue and what began as being an exercise to strengthen a weak hive has now resulted in just killing the majority of your bees off. The method I like to use is a dose of water. The water should be tepid - too cold and it will shock the bees - and mist it over the bees with a sprayer that you get from a DIY shop or garden centre. Make sure, though, that if you are not using a new one it is thoroughly washed, as any sort of chemical residue will easily kill the bees. During my early days as a bee keeper I learnt the hard way. I once used a sprayer that I had used before and assumed that it was clean. I filled it with tepid water and set about spraying the bees. What I didn't realise was that the last liquid in the sprayer had been ammonia for killing off red mite in the chicken coop. Bye-bye bees. Bees may be hardy little creatures but even they have their limits.

So there you are spraying your bees with tepid water. So what's next? The next stage is to place the donor frame complete with the brood and its adhering bees into the brood nest of the receiving hive. Open up the receiving hive. Make a gap in the centre and give the gap a good smoking to drive the bees away. Spray some water into the gap and drop the frame in gently. Whilst the bees are busy with their impromptu bath you can close everything up.

There is another way of uniting a colony and one that never ceases to amaze me. This method is usually done when there is more honey in the hive, however. Essentially you move the bees and the supers from a very strong hive to a medium strength one. Don't try it with a weak colony as they will not be able to look after the honey properly, which will promote robbing and the eventual destruction of the colony. So choose a strong colony and a medium colony to marry together. You must bear in mind that the uniting of two colonies is a completely unnatural process. Bees are territorial creatures and are a little unkind to bees from other colonies that might stray a little too close. (Two swarms can unite, however, but this is far from the norm.) Bees recognise each other through a chemical signature. When you unite a hive clearly you can't duplicate the signature, in fact you have two separate ones and this can lead to fighting amongst the bees. Of paramount importance is to remove the weaker of the

two queens - you can't have two queens in one hive. The way I like to bring two colonies together is through a 'paper marriage'. This method has always been seen as the safest way to complete the job. The best time of day to set about performing the marriage is at dusk. This is because the bees will be back in bed for the night or will at least be flying less and will certainly be moving more slowly.

Tips!

- **If your brood is weak, check for disease first.**

- **Always use clean instruments.**

- **Dusting the bees with icing sugar will keep them occupied and help prevent any fighting between the receiving and donor bees.**

If you are working on a Langstroth hive, which is the one I like to use, remove the roof and split the hive by taking off the super and placing it to one side. Cover the top of the bars in the brood box with a single sheet of newspaper, preferably a good quality broadsheet that will stimulate the bees into reading. Make a few holes in the paper with a small nail or two cocktail sticks taped together. Make sure the holes are not too big, though, otherwise the bees will be able to push their way through immediately and fight with the other bees. Take the brood box from the other hive and place it on top of the newspaper. Line the box up accurately and make sure there are no gaps. The bees will then pick up the scent of the other bees and will try to work towards them. You will have a very good idea that the exercise has been a success when you see shreds of newspaper on the alighting board and if your bees are now wearing spectacles. At this stage the bees that have been introduced have lost their old queen's chemical signature and picked up the new queen's signature and have therefore been accepted into the hive. Leave this new colony for at least a week before doing anything with it.

APRIL

My Grandfather used to keep bees and always used to say a few lines from a poem he knew as a child: 'When daisies pied and violets blue and lady-smocks all silver white, and cuckoo-buds of yellow hue, do paint the meadows with delight'. I remember keeping bees with him as a child and he painted a picture of the bee keeping season to come with all the colours and the expectations of a good harvest. And now I always feel that this is the part of the year when it all starts to take shape. You may find that bad weather in March has prevented you from cracking on with your work schedule and you are playing 'catch up'. But let's work on the basis that we have had a sharp winter and March has been a good month, with decent weather and cool nights. April then is going to be even better; the clocks have gone forward, the trees are in leaf and the bees are more and more active. Around the middle of April you should find your bees expanding - not the actual bees, of course, but the colony. This happens with the arrival of much more available pollen and an increase in nectar.

If you find that you have been bitten by the bee keeping bug and now have hives at several locations, you may wish to take advantage of swapping bees around on good flying days. These are days when there is a good flow of pollen and nectar, the weather is right and the colony is increasing at a good rate. If you have a colony at another location that is struggling a little, swap it with a stronger one. At any time close the hive up (see March section) and move it to the exact location of the other hive. The bees that were away from the hive in full flight will settle, awaiting a replacement hive which, when it arrives, they will cling to and eventually enter. In theory the bees from your successful, strong hive will enter your weaker hive and give you two medium strength hives. It is advisable to re-queen what was considered initially to be the weaker hive. The stronger queen will lay eggs quickly and will therefore take the hive back up to strength fairly swiftly. ALWAYS CHECK FOR DISEASE BEFORE COMPLETING ANY SWAPPING. I do like it when a plan comes together. Tea time, I think.

At this time of the year it's wise to check for any damage to hives, frames and location. Never work on the basis that everything will be all right, as it seldom is. You may wish to keep a small diary to jot down your findings and to compare what you were doing at the same time the previous year. You can note everything from the weather to tricks you have picked up that you may feel will be of use next year or even where you have sited your hives. As well as keeping a general diary it is also a good idea to keep a separate record card for each of your hives. You can either buy these or you can compile your own. Buying some is the easiest as they will take account of items that you will probably forget to include. You can get some from either a bee keeping club or a bee keeping

equipment supplier. On each card you should include information such as your hive number, where it has been sited and moved from and to. You should also note down the number of the queen, her colour and breed. In addition, if you have treated your hive for disease or mite infestation, record the type of treatment, batch number and when you carried it out. It's useful to include the contact number of your bee inspector. The more information you document, the better, as your bee inspector will visit you during the year and it will make his/her job that little but easier if you have all your information to hand. It would be wise to use the card to note how much honey the hive has produced. This, combined with all the other information, will be useful in enabling you to make any necessary changes to the hive when the next bee season starts.

If the weather on Planet Craig is still going to plan you should find the hive is building up quite nicely and the queen is busy egg laying. Just be careful that overcrowding may be taking place. When I first started with bee keeping my greatest fear was of swarming. I had a vision of a ball of mad and angry bees invading someone's central heating system or 100 feet up a tree. My vision was unfounded as you can cope with a swarm quite easily and neatly. Bees at swarm are unusually quiet, if not docile. If you have quiet, non-aggressive bees anyway, then when they are swarming they are laid back almost to the point of falling over. Even my feisty bees at swarm are very easy to deal with. Before today I have collected swarming Iberica bees from a tree with my bare hands without any protective clothing and have placed them into a cardboard box and carried them home. My mother used to say that they were bees you could invite around for tea. But then my mother was a little eccentric to say the least.

If your bees are becoming overcrowded, then their thoughts will turn to swarming. 'Don't panic, Mr Mainwaring!' If you have the equipment to hand all will be fine. (See section on swarming) What you have to remember with overcrowding is that there is no space for the queen to lay eggs, there are no empty cells and there is no room to store pollen or nectar. To alleviate the problem simply take some of the frames from the full brood box and replace them with empty ones. If you do take frames out, remember to take only those with store food on them and not eggs, and always leave some full frames. You may find, for example, that by taking three frames out of a ten frame chamber you will have the breathing space you - or rather the bees - need. You may also wish to add a super at this stage to allow a certain amount of freedom within the hive. The frames you have removed with store on them won't be lost, of course. You will be extracting the honey from these later.

We have not spoken about the extraction of honey yet and although I will cover this in a separate section it is worth mentioning one important job you can do in preparation for later in the season: sterilise your extractor.

MAY

Ah, on Planet Craig it is warm with a slight breeze and not a cloud in the sky. Spring is a distant memory and summer is here and ready to be celebrated. Reality can be somewhat different, but welcome to my world.

Late April to mid July is swarming time and this is mainly dealt with in the section on swarming. A tip to consider at this time of year, however, is to have one or more 'bait' hives ready for any swarms. I can guarantee that you can check your bees on a regular basis, looking for queen cells and the possibility of swarming, only to be contacted by an irate neighbour as soon as you have just sat down for dinner, complaining of a swarm of bees in their garden. If you have a bait hive ready and waiting it might at least attract any errant swarms that appear.

Apart from dealing with swarms what should I be doing now? If all is going well your hive should be jam packed with bees and you should be thinking about placing a queen excluder on top of the brood box with a super on top of that. The queen excluder is simply that. It is either a metal or plastic frame that excludes the queen from getting into the super and laying eggs. The super should be free of eggs and used exclusively for the laying down of honey. Keeping the queen away from the super also makes it easier to extract the honey at a later stage as it is not contaminated with eggs and bits of bee.

As you start to replace hive parts, especially frames, make sure the wax you are using is clean and as fresh as possible. If you are storing wax from one year to the next, try to keep it tightly enclosed in a plastic bag. The reason for this is to keep the wax moth away (see section on pests). Wax moths will lay their eggs quite happily in little nooks and crannies in the hive, but mainly in the frames. The bees will work quite well alongside the wax moth, but when the larvae reach a critical stage, they will erupt and eat all the available wax. An annoying habit of the larvae is that they are happy to lay dormant in a corner until the time is right to emerge. You can easily tell if you have

Tips!

- *In the event of overcrowding, simply take out some of the frames from the full brood box and replace them with empty ones.*

- *In preparation for honey extraction, sterilise your extractor early.*

- *Late April to mid July, have a couple of 'bait' hives ready for any swarms.*

wax moth larvae as there will be a trail of black and white gritty, sand-like deposits in a line, or you will see a line through the wax indicating where they have gone on their holidays by eating through the wax. If too much wax is eaten it can spell disaster for your colony. How can I avoid the dreaded wax moth, I hear you ask. Well, one of the best ways, apart from keeping the wax in sealed plastic bags, is to take your wax, either in sheets or ready-framed, to a local cold storage depot. For a pro-rata charge of around £50 per pallet they will give you a quiet corner where you can leave your wax in sub zero temperatures, so if you are only taking a couple of bags or frames then the charge will be incredibly small. By doing this you will effectively freeze all the eggs, larvae and adults which will kill them off. It is one of the most natural and effective ways to eradicate this pest.

Another job for this time of year is to change over your frames. You can be certain that by putting new frames complete with new

Swarm

if you are unfortunate enough to suffer disease in the colony. In order to moderate any impact from disease it is advisable to replace about 35% of the frames each year so that, in effect, you have a brood box that is no more than three years old.

Frames containing old food should be removed as a matter of course. Leaving old food in the hive promotes robbing and encourages other insects to make a home in your colony. You can return these frames to the hive in the autumn to be used as winter food. In the meantime, store the frames in the freezer which will kill off any bugs. We know there is a lot of goodness in honey for humans but honey is primarily bee food, just as a cow's milk is intended to feed calves. The many and varied properties in honey are formulated to keep bees alive and well. It is important, therefore, to give the bees as much of their own honey as possible, even if it means a reduced honey harvest.

wax into the hive they are clean and safe. An added tip is, if you are replacing a frame in the brood box, mark the top bar with the year - you will then at a glance be able to tell how old the frame is. (There is no need to mark frames you put in supers as you will be removing these anyway when you harvest the honey.) Some bee keepers believe there is no need to mark their frames and several have told me that they can identify pre-war frames. I believe, however, that it is good practice to mark the frame, particularly

Before the honey does start to flow and you, naturally, get all excited, remove any varroa treatment, whether it is strips, wads or pads, and place them to one side as the chemicals found within the treatment will taint your honey. More on varroa treatment later.

It is helpful to have an empty brood box available. This is useful for carrying empty and full frames to and from the hive. Equally, it's extremely useful to have spare super boxes made up and to hand. (See March section on framing)

Super frame (shallow)

On one of your routine inspections you may open your hive only to find other insects as well as bees present. The only other insect that you should feel happy to have as an uninvited guest is the common or garden spider. This is a good sign because it indicates

that the colony is warm, dry and draught free. Spiders, like bees, love warm, dry conditions and hate draughts. If you have ants, move the hive a distance away as once the ants find a readily available supply of sugar in the shape of honey, they won't leave until they have removed every last ounce. Earwigs will try to make a home in the top of the hive as this is one of the warmest parts. Kill them off as quickly as you can as, although they are no direct threat, they can carry disease. (See section on disease)

JUNe

If the weather is warming up and it's a flaming June (I can but wish) then it's important to ensure that your bees are properly watered. Without a good supply of fresh, clean water your colony can quickly be in trouble. The supply of water doesn't have to be right next to the hive, but try to ensure that there is a natural or man-made source (stream, canal, garden pond, for example) nearby. If, however, the only source of water is some way away, it is not a good idea to make your bees fly long distances on a hot day to reach it. In this case leave a shallow basin of water near the hive with some flat stones in it so that the water meniscus is broken and the bees are able to stand on the stones to access the water without drowning. Be sure to top up the water regularly. Water evaporates remarkably quickly in hot weather and other creatures may well rely on it too.

A good supply of water isn't just essential for the bees to drink. If it gets very hot outside the internal temperature of the hive will increase accordingly, especially if you are using a polystyrene hive. The bees will need

a constant supply of water to air-condition the hive. The bees will draw a droplet of water from the source, put it on their backs in between their wings and return to the hive. They will then fan their wings and disperse the water, thereby cooling the hive. The more time they have to spend collecting water to reduce the hive temperature, the less time they have to make honey for you.

Improvised water trough

As a good bee keeper you must continue to stay on top of your disease control. Don't get me wrong, bees don't come down with disease at the drop of a hat. But good bee husbandry will help to prevent disease rather than having to deal with it once it has taken hold. If you do find something about which you are unsure then contact your local bee inspector and ask for their advice. They are always keen to help and whether you have 1 hive or 1000 hives, they will always help and it's better to be safe than sorry.

By June you should find that the brood will fill most of the brood chamber. Don't be too worried if the bees have not managed to fill the brood completely, but if you find that the bees are only at three or five bars then you may want to consider marrying the weaker hive with a stronger one rather than lose a colony altogether (see March section). I prefer to work on that old Irish saying my father used to come out with: 'Better to have half a loaf than no loaf at all'. But if your bees are not performing well and increasing at the expected rate, check for disease before assuming you have a weak queen. Often, however, it's better to check for disease and re-queen just to be on the safe side. By your second year you should have a supply of queens.

Sometimes June on Planet Craig isn't as warm as I would like. It can be an odd time in the Northern hemisphere for bees; indeed in the United Kingdom, it's referred to as the June gap. This is the time when the spring flowers that have given a boost to the bees have gone and the summer flowers have yet to make an appearance. You can beat this by planting up appropriate flowers to fill the gap in June (see chapter on gardening for bees), or you can give them a little fondant or syrup feed. Don't give them too much, though, as it can encourage them to become lazy.

Remember, if you go to your bees one day and then revisit a couple of days later, wash your gloves and bee suit between visits to help to prevent any cross contamination of your colonies. (This is why I like to use disposable gloves.) The same rule applies if you have visited a friend's hive. You should also sanitize your bee keeping tools in a sterilising fluid.

I am often asked how is it that you are so good looking ...? And then I wake up and realise the exact question was how often should I visit my hive. Too much attention, vibration and movement will upset the bees, but leaving them alone for too long may make them angry when you do go to pay a visit as they will have become used to a very quiet life. Bees have a poor memory. Don't ever think the bees will remember you. They will respond to being handled gently and carefully. But they won't thank you for being dropped or shouted at - even a conversation taking place between you and someone else could upset the bees. My rule of thumb is to work with them for an hour a week between April and September and an hour a month thereafter. As long as you are being gentle with them you could visit them every day. Remember, however, that bees have to maintain a constant temperature in the hive and if you frequently open it up then they have to work much harder to preserve that warmth. More work means more energy which in turn means more food. And in the summer bee food means honey. The more honey the bees consume, the less honey there is for you.

JULY/AUGUST

This is my favourite time of the year, certainly in terms of bee keeping. There is a feeling of being established by July: you know that frosts will not hamper your efforts and that flooding isn't going to wash away your hard work. I was once in New Zealand, on the South Island at a little village called Ross, population 83. I met an elderly couple who produced honey. They had scaled down from 900 hives to 300 as they had grown older. They were genuinely pleased to meet other bee keepers and after several hours

of tasting honey and chatting they offered my wife and I the opportunity to buy their apiary. It looked great but I was concerned about knowing very little about their climate, forage for the bees and so on. In addition to that there was only one shop, the nearest cinema was four hours away and the year before 300 of his hives had been washed away in a flood. That sealed it for me - we said no. At least in the majority of towns and cities flooding isn't generally a problem.

Well you should be at the stage where swarming is generally over. There is an old proverb going back to 1655 that says: 'A swarm in May is worth a load of hey, a swarm in June is worth a silver spoon, yet a swarm in July isn't worth a fly.' Essentially it is saying that in late spring (May) a swarm is a good sign, a swarm in June is a very good sign indeed but a swarm in July....well it's not really worth keeping. It will be late, weak and weedy and unlikely to survive over winter. You would be better to destroy the queen and incorporate the bees in a weaker hive. This proverb may be hundreds of years old but it still rings true.

You should be at the stage where your brood box is full, your supers are on and your colony is spending its time looking for nectar. You may read of honey being raw. This means that the honey isn't capped. Leave it in this state in the hive until the bees have capped it and once this has happened, remove it and extract it. Once it has been extracted (see extraction section) the frames are considered to be wet. Put them back in the hive and the bees will eat the honey that hasn't been extracted and will clean and dry the cells out ready to be filled with honey again. If for some reason you can't replace them wet, put them to one side to be dealt with soon after.

You may be lucky enough to have a hilly district close by on which you will find a heather moor. If so, make the most of this golden opportunity by finding out who owns the land and asking them if you can put your hive/s on it when the heather is in bloom, which is usually around the second to third week of August through to the early part of September. You should now think about treating your brood for varroa. As I said earlier, you don't want to contaminate your honey with chemicals, so you should treat the bees for at least a cycle of the bee and a bit more before you move them. As a rule of thumb I like to treat the bees for around a month, so in order to move the bees in time for the heather you should start your treatment at around mid July. You should also ensure that you have a current year queen to make sure the brood production is kept going. At this stage you should remove the queen excluder and any excess honey. The hive must still have plenty of store food, however, to ensure that the bees are fed until the heather honey starts to come in. You should also put in some wet frames that you kept from earlier into the supers as the bees will need drawn comb to give them a head start. Remember too that it's a lot colder on the moors and although heather will give off nectar even when it's cold, you might find that the bees are disinclined to leave the warmth of the hive. (The crucial temperature for individual bees is 6°C - at and below this they will become inert and die.) Heather produces a beautiful, dark almost caramel-type honey; it's thixiotropic which means that it neither stays runny nor does it set completely. And because of its distinct pedigree it needs to be extracted in a totally different way to other honey: this will be covered in the extraction section.

Ivy is commonly found in urban areas

If you do not have the opportunity to move your bees to some heather do not worry - there are many plants that flower at this time of year in towns and cities. (See the chapter on gardening for bees.) One of the plants that I look for in the towns and cities is ivy. Ivy really attracts bees in their droves as it offers a good strong supply of nectar.

You may find that if there is a good flow of honey the hive is plagued by wasps. They are intent on robbing honey from the hive and will attack the colony in order to get at it. Remember, worker bees have one sting - wasps have multiple stings. One way of keeping wasps at bay is to place a bee guard over the alighting board. This is a bit like a mouse guard but with smaller apertures - it will allow access for the bees but not for the wasps. You can also reduce the hive entrance by about 75% by using a piece of foam rubber (See March section). This will allow the bees on duty to patrol a much shorter distance, keep unwelcome guests at bay and will also allow you the opportunity to identify the strength of the hive by the number clambering to get in.

When your flow of honey is complete and/or your frames are capped, remove the frames and place to one side. Start extracting as soon as you can because the natural heat from the hive will ensure that the honey flows quickly without having to artificially heat it, thereby losing some of its beneficial properties. If you cannot extract the honey straight away, make sure you leave the frames somewhere where vermin or wasps cannot get to them.

Well, congratulations! You have just finished your first successful year as a bee keeper! You will have attracted lots of new friends desperate to know about the powerful uses of the golden liquid. You will be asked many and varied questions and I have to admit that some of the questions I have been asked over the years have flummoxed me. I once had a lady at a Women's Institute event who raised her hand in front of a full meeting and asked if honey can be of any use for an inactive sex life. I blushed, (wondering exactly how the honey might be administered) and explained that I had never known it to be of particular use in this area. She then blushed and said it was for her husband and not for her. Honey does have very good medicinal practices but this is not what my book is about.

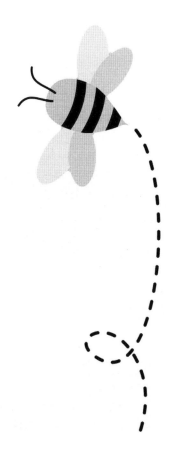

Chapter Two
PESTS AND DISEASES

I want to concentrate on the areas that are of utmost importance to the bee keeper. These are not necessarily humorous or flippant, rather they are integral and vital sections that cannot be left out. Bee keeping has always been referred to as the science of agriculture. I can not stress enough that apart from being your own vet, you need to be your own weatherman, scientist, woodsman and most of all wise old sage as you will be asked a thousand and one questions about the subject. Perhaps the most important subject, however, that the bee keeper must be vigilant towards (and I have stressed this from the outset) is pests and diseases.

Varroa Mite

Let's first look at some of the more common pests and diseases that you might come across. We'll start with ones you may even have heard of, but don't worry too much as the majority of them are treatable. Firstly, varroa mites.

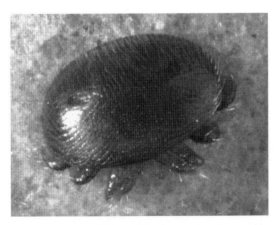

Varroa mite

Varroa destructor and its relative Varroa jacobsoni are both parasitic mites that spend their time feeding off the body fluids of their hosts, these being mainly the adult bee, pupa and larva. If you are looking to see if there are any mites inhabiting your colony there is no need to buy a microscope to do this as the varroa mite can be seen easily with the naked eye. They look like fast-moving pin heads. If you see them running around the hive and not on the bee it doesn't mean they are moving out, it means you have a bad infestation and therefore there are too many mites and not enough bees and the mites are running around looking for a host. They are a red-brown colour and will be found around the bees' thorax. The varroa doesn't make life too comfortable for the bees: it's like a human having three dinner plate-sized bugs on his/her body sucking the life out them. Not nice. It's the virus that the mite carries that really does the damage, however. The virus with which the bees are infected can be identified when the bee erupts from its cell. The bee will invariably have some form of deformity. Often it is wingless, sometimes only half a bee emerges, or has legs missing, is weak or blind. As long as your hive is relatively fit and healthy it can withstand some varroa invasion. A crucial time is when the colony starts to prepare for winter; you may find that the mite population increases leaving the bees' growth behind. It is at this stage that the bees may appear to swarm but in reality have left the hive to seek shelter elsewhere. If the infestation is very bad the colony may even die. Colony Collapse Disorder (CCD) has been associated with the virus carried by varroa mites. More on CCD later.

What can we do about it, I hear you cry! The good news is that because varroa is a worldwide problem, treatment is readily available. One option, but arguably not the best, is to do nothing. This might seem a strange choice but I have read a number of pieces written by a range of well established bee keepers that draws on the issues associated with natural selection. The theory is that if you leave the varroa to live with the bees for long enough, they will become good friends and eventually invite each other for tea. Personally I think this an ill-advised and dangerous approach. The Australians, however, are trying to breed a queen that lives alongside the varroa mite and it appears that some trials have been successful - so watch this space. Luckily, there are other chemical and non-chemical treatments available.

Female varroa mite

Firstly the non-chemical approach. This is ideal for those bee keepers trying to produce an organic honey who do not want to include a chemical element. This approach generally relies on the disruption of the mite's life cycle and is unlikely to eradicate the mite altogether. If you can keep the number of mites down to a relatively small quantity, then they won't be too much of a problem.

So let's look at how this can be done. One way is called the 'drone brood sacrifice', which taps into the preference of the varroa mite for the drone brood over and above other brood. Remove a frame that has drone brood in it and place it in the deep freezer overnight. Then de-cap the comb and rinse out with fresh water Shake any excess water from the frame and return it to the brood. If all goes well the queen will re-lay in those empty cells. Don't remove all the drone brood frames at once, but you can remove all those with drone brood on them. Started early enough in the season you can control the spread of the mite.

Some bee keepers extend the 'drone brood sacrifice' to capped worker brood too. In my

Tips!

- *Dusting the bees with icing sugar is a good natural remedy for dealing with varroa mites.*

- *A mesh floor offers more in the way of air circulation, which helps prevent disease.*

opinion, however, this is economically foolish as the amount of worker brood sacrificed in relation to the amount of varroa mites removed rarely even balances. Forfeiting workers will always leave a weak brood and to me this will consistently spell danger. You can, moreover, trap the queen on worker brood, which will also limit the amount of area that the mites have to reproduce in. You can then remove a sealed worker brood and deal with it in the same way as the drone brood. You can do this three times but you have to be very careful with the timing. Factors such as the flow of honey and the weather can jeopardise the success of this method because the loss of brood could deplete the colony to such a point from which it might not recover.

I have found these approaches to be time consuming and incredibly labour intensive and, because you are constantly interfering with them, the bees are put under a great deal of additional stress.

Another non-chemical method, and one of the most pleasant ways to dislodge the mites I have tried, is to dust them lightly with icing sugar. It may seem strange but once you have tried it and you see that it is effective, you may wish to continue throughout the year. Firstly gently remove the top of the hive and allow the bees to make their way up to see what is going on. Just before they decide to fly (this is the crucial point), dust them with icing sugar using something with a fairly fine mesh - the holes in a colander may be too big, but a sieve that is used for taking

Bee with varroa mite

the lumps out of flour will do. Like monkeys at the zoo looking for salt on each others' coats, the bees will pick at each other to try to dislodge the icing sugar and at the same time they manage to knock the varroa mites off. Hey presto - mite-free bees. This is only successful, however, if the floor of your hive is mesh and not solid. Once out of the hive, the mites cannot crawl back in or 'hitch a lift' from a bee returning to the hive. Until fairly recently most hives had solid floors. It was thought that this would reduce draughts and therefore keep the bees warmer, especially during the winter. This may well have been true, but a mesh floor offers more in the way of air circulation, which helps prevent disease, and it is far better from a varroa-control point of view.

Secondly, chemical options and, more importantly, do they work? Well, done properly and at the right time you can have great success with a variety of chemical applications. These applications essentially choke the mites as they breathe through little trumpet like tubes on their backs - but please don't feel sorry for them at all. As the chemicals work on the mites, they choke, fall off the bee and you can wave goodbye to them. You do need to leave the chemicals in the hive, some people say 21 days, others say 42 days. 42 days is a luxury that most people can't afford, especially if you have a good flow of honey and you want to move the hive. I generally work on 21 days plus a few. This gives time for all the bees to be born after the initial application and for the mites to have been removed. You will find there are a number of products available on the market. Thymol is marketed as Apiguard or Apilive; it is effective but seen as a 'soft' chemical. Fluvalinate is marketed as Apistan and coumaphos as Checkmite. I have used Apistan which comes in the form of plastic strips, usually five to a bag, so enough for a hive. If you distribute the strips across the brood then the bees will brush over the strips as they follow the queen around, causing the chemical in the strip to adhere to the varroa mite. It's a pretty effective treatment. And the down side? Well, removing the strips from the bag means you have to take off your large, safe, bee-proof gloves and if your bees are a little aggressive or you are not a confident bee keeper, you will get stung. Another down side I find is that once you have placed them over the frames and left them for the allotted period, by the time you return they can be gummed up with wax and propolis. Then, as you try to remove them they snap and fall to the floor of the hive which leaves you having to retrieve them. And as we know bees resent being disturbed too much. These treatments are quite effective although the mites, like many other parasites, have started to become tolerant to some formerly effective chemicals.

Tips!

- **If using chemicals to clear mites, work down wind or you'll end up with a head-ache!**

- **The crucial period of growth for the small hive beetle is at the pupation stage which takes place in the ground immediately outside the hive. This is the time when the pest can best be destroyed.**

- **Acarine mites are probably better off getting the bus!**

Some bee keepers will recommend the use of 'natural' acids such as oxalic acid or formic acid, but take care with these as neither is registered for use in the UK. You can, however, use a preparation called Miteaway which contains formic acid. This comes in a gel pack and seems pretty effective. The packs come in a plastic bag inside a sealed bucket wrapped in plastic. I always thought that was strange until I got a whiff of the contents - I quickly developed the mother of all headaches.

Inspecting the Drone Brood for Varroa Mites

Once out of the bag, place the pad on top of the frames. The pad sits on little runners that can either be bought or made from sticks. The gap allows the bees to get under the pad as well as over it. The results I have found have always been very good but there are negative points to bear in mind. Firstly, the chemicals do give you a headache, so work down wind and secondly, if you don't get the spacing right and too much pressure is put on top of the hive roof, you effectively squeeze any of the chemicals out on top of the bees, shortening the life of the pad and overdosing the bees on formic acid.

Acarine Mite

People talk about mites in bee keeping circles and more often than not they are referring to the varroa mite. Be careful when the word mite is used because it could refer to another one called the acarine (or tracheal) mite. This is a tiny mite, not visible to the human eye like the varroa mite, so there is no point in looking for it unless you have a microscope. It is a parasitic mite that infests the honey bees' airways. You may come across it with reference to the Isle of Wight Disease in the early part of the twentieth century. This disease was found in 1904 but not identified until 1921 and in that time it had annihilated almost the entire bee population of the United Kingdom. Since then a resistant bee has been bred and is now widely available. These bees are known as Buckfast bees and were specially bred by Brother Adam at the Buckfast Abbey. The only way to detect acarine mites is through microscopic inspection and dissection of a sample of the bees. When the female mites reach maturity and are impregnated they will exit the bee's airway and relocate on the hair of the bee. They then either wait for a number 12 bus or a passing young bee. Usually a bee comes first and once on the young bee the mite will re-enter the airways and start the process all over again.

Treatment is simple. All you need to do is combine 25% vegetable fat and 75% icing sugar to make a paste and place it on the top bars of the hive. The bees are keen to eat the sugar and whilst they are munching away they pick up the fat with it. We are not sure how, but this rather cunningly disrupts the acarine mites' ability to successfully identify a young bee. The mites become confused and whilst some will successfully detect a young bee, many more will remain with their original host and others will transfer to another older bee, which may already be infested with acarine mites. A larger proportion of mites will die before they have the chance to reproduce than will successfully produce another generation. Clearly the mite should have taken the number 12 bus.

Another easy way of treating a hive contaminated with acarine is to use a simple application of menthol. This will certainly kill off the mites. You can apply the menthol directly as a crystal and allow it to vaporise in to the hive, or if you mix it into a carrier medium of soft bees wax it can then be placed on the top bars of the brood frames.

Small Hive Beetle

Small hive beetle is not really a problem in the United Kingdom. The beetle is a small, dark coloured beetle that enjoys nothing better than living in a hive. As this book is scheduled not just for urban bee keepers in the UK but worldwide, I will explain that small hive beetle was originally found in Africa, where it was happy to spend time being a pest. Then with a huge leap and a bounce it appeared in Florida in 1998, and by 2007 was as far north as New Jersey and as far west as the Pacific coast. The crucial period of growth for this fun loving criminal is at the pupation stage, which actually takes place in the ground immediately outside the hive. This is the time when the pest can best be destroyed.

Lesser wax moth

One approach that has been recommended by American bee keeping societies is to use diatomaceous earth. This is a naturally occurring, soft, siliceous sedimentary rock that is easily crumbled into a fine powder. What you do is to spread this powder around the hive. As the pupae emerge, the earth acts as an abrasive, scratching the surface and causing lesions. The bugs then dehydrate and die.

Greater wax moth

Three cheers for diatomaceous earth. Its original use was to prevent ants from entering the hive and robbing your precious honey, but it also appears to be effective in reducing hive beetle numbers. Diatomaceous earth is used in cat litter and some bee keepers do use cat litter around hives to prevent problems with ants. There is a secondary issue, in that it encourages cats to poo outside your hives. I have also found that dogs, and especially their cousins, foxes, are attracted to the cat litter. And where you have a fox there are often badgers around. Badgers will attack a hive if they are very hungry, so you have to decide on the greater or lesser of two or several evils. I would always opt for getting rid of the small hive beetle and putting up with Mr Fox and his countryside fellowship.

There are also approved pesticides available to use within the hive. These are aimed at the adult beetle. What you do is to place the prescribed amount of chemical in the grooves of some corrugated cardboard. Use a standard or large gauge cardboard to allow the mites to enter freely into the cardboard - make sure the grooves aren't too large,

though, because the idea is to let the beetles in but to keep bees out. The beetles like a nice warm, dark environment - I can think of places like parts of Yorkshire that would suit - to munch away in, so something like corrugated cardboard will be like inviting them for a posh, free meal - they won't be able to keep away.

Wax Moths

Ahhhh! Planet Craig's least favourite and most painful visitor! I need a cup of tea before I go any further. I have such problems with these little pests here in Lancashire. If you live in a temperate climate, somewhere where there are warm winds, rain and occasional foggy days, limited winter snow fall and so on, much like Planet Craig, then that's the place where wax moths like to live. They are not the best of dinner guests; they are bad mannered and have poor table and personal habits. I remember when I was a young chef, a mere 15 years old, my first head chef said to me: 'we are here to feed not fatten'. I have worked with this until I came out of catering and went into law. The life of the bee keeper is really no different. We are here to manage the bees and support the environment and if we are lucky, we get some nice sticky honey at the end of the season. We are not here to provide a halfway house for some degenerate moths who think they are on to a free ride and meal courtesy of a woolly-headed Lancastrian.

The greater wax moth or larger of the two species, is moderately sociable, however, and will live side by side fairly harmoniously with the bees, munching its way through the wax made for the honeycomb. But that's where social niceties end. The large wax moth's development requires access to used brood comb, or brood cell sweepings, which may be found either in corners of the hive or frames, in nice little groves under the top bar, or just about anywhere dry and warm. (Florida would be ideal for wax moths.) The reason why the moths need the wax is because it contains proteins essential for the baby wax moth to develop. The larvae develop in the form of brood cocoon. As the young enlarge they will split the comb and this will cause the honey to spill out and in turn may kill the baby bees. You may find that in warmer parts of the world, when honey is stored in supers and hasn't been removed from the comb, any surviving larvae can easily destroy the entire comb. You can see the trail left by the larvae as it works its way across your expensive wax frames.

If you do detect wax moth damage in your hive, you can scrape out the damaged comb and have a quick check for any stowaways. Don't worry about the damage to the comb as the bees will repair it and make it good. If you have a strong, stress-free hive, the bees should stay on top of the moths for you. The bees will generally remove any larvae of the large wax moth developing in the cells, but the smaller wax moth will happily lay its eggs in any cracks and crevices and will generally go undetected by the bees until they become fully fledged moths. Fortunately, there are a number of ways that the bee keeper can control these moths. A control method for stored comb is to introduce the 'aizawai' variety of Bacillus thuringiensis: this is a bacteria that can be sprayed onto the comb with a plant humidifier. You have to give a good coverage, however, because the bacteria must be eaten by the moths to be an effective control. Be wary of being offered other chemical controls, however, as although they may be very successful in dealing with wax moth, they can also harm your bees. I am thinking of paradichlorobenzene and naphthalene in particular, which are active ingredients in domestic mothballs.

I am always keen to use natural methods of eliminating bugs and one of the most effective ways of dealing with the wax moth is to expose them to freezing temperatures for at least 24 hours. If you live in an area that regularly has hard winters with temperatures to match you can leave your frames, supers and so on outside in a dry but open space where they will be subjected to a good, hard frost. If, however, you enjoy a balmier climate and the temperature is unlikely to drop below freezing, then a deep freezer will be an ideal substitute. (See May section)

Nosema

One of my scariest times in bee keeping was when I was a novice. I was a young man, keen to be the best bee keeper in the world. When spring came I was like a coiled spring ready to leap into action. It was a warm day in April and I felt great as the bees started to come out and fly. They would land and leave several drops of a yellow liquid, like a weak, turmeric-coloured tear drop. In my innocence I thought they had found a supply of yummy pollen. In fact they had nosema, otherwise known as the bee runs. As you may know, dysentery in humans can be a killer. And it can be the same for the bee and its workmates. You may find that nosema becomes a real problem when we have a long spell of winter weather and the bees are confined to the hive. If they are cooped up and, more importantly, short of fresh air, then the trouble can start. If the weather is such that they manage to fly and evacuate themselves, then that generally cures them.

Prevention is better than cure, and although increased ventilation isn't a 100% foolproof method of preventing the disease, it will go a long way towards it. Increasing the ventilation can be achieved by not just fitting a mesh floor (these tend to be standard these days), but by standing the hive on either three old tyres or an old wooden pallet. This will not only allow more air to circulate but it will also prevent dampness working its way up from the ground.

If the nosema is severe you may have to treat your bees with antibiotics which are introduced via their food. Essentially you need to warm up some syrup and add the antibiotics. The bees will be attracted to the warmth of the food, especially at the time of the year when nosema is more prevalent. But that isn't the bee all and end all. There is an ongoing debate that you can minimise the effects of nosema (you're not going to totally eliminate it) by removing all the available honey from the colony and feeding the colony on a syrup solution of one part water to one part sugar. The water and sugar should be boiled together until the solution reduces. Although using cane sugar can be quite expensive, it is generally best because it contains less ash, which reduces the risk of dysentery, and the bees certainly seem happy with it. The debate rages between bee keepers as to whether it is better to take the 'home-made' approach and make your own syrup, or the commercial one of fondant or ready-prepared syrup. Personally, I like to use a commercial syrup as it generally contains specially formulated additives that support or enhance the colony - a bit like giving yourself a tonic. The other side of the debate advocates leaving the honey in the hive, since honey is the most natural bee food of all and therefore must be the better option. Not so for the bee keeper, of course. After all that work through the summer what you want is a nice bit of honey!

Foul Brood

Just when you thought you've done all you can to prevent your charges from getting any nasties, along come two that will upset you the most. These are American Foul Brood (AFB) and European Foul Brood (EFB)

American Foul Brood

One of the most upsetting times for any bee keeper is to discover you have a diseased hive. It is soul destroying. I have always thought myself to be a diligent and orderly bee keeper. But when the regional bee inspector popped in to have a look at my hives, his discovery was a complete shock. I had no concerns and invited him to have a routine look and I would carry on the idyllic life of the apiarist. Not to my surprise, but to

my shock and horror, AFB was present in my hives. In fact it was present in only three hives. I had these hives on a rape seed crop and they were just ready to be moved on to field beans. I was upset to discover that the bees were now subject to a stand-still order. This meant that I was unable to move any of them until the inspector had received information back from the inspectorate to give me the all-clear. This stand-still order demands that bees are not moved for six weeks minimum, or until permission to move

American Foul Brood infected comb

them has been given by the inspector. The inspector took away the contaminated frames for inspection and said he would report back to me as soon as possible. If the inspector identifies what he suspects is AFB or any other notifiable disease, he will order the colony, complete with hive, to be destroyed. I had to destroy three hives. It broke my heart. I initially felt like I was a bad bee keeper. I felt dirty. What had I done, or hadn't done, to introduce or encourage such a disease? Apparently, I felt the same way that most bee keepers do when they are told this. But you should not. All you need is a rogue drone which has visited an infected hive to bring it back to one of your hives and pass on the virus. It is as simple as that, and cannot be foreseen by any bee keeper. Fortunately, after six weeks I was given the all clear and my hives were allowed to be moved. Once you have had a batch of disease you never forget it, though.

If you are unfortunate enough to have to destroy one of your hives, it has to be done in the presence of the bee inspector. You have to dig a pit at least three feet deep and put the hive into it. First you have to kill the colony. I was initially told to drench it with petrol and allow it to soak in. I found this extremely upsetting as the bees seem to scream when the petrol is poured on them. An alternative is to use lots of warm, very soapy water. This blocks their breathing vents and they suffocate. It is a lot more humane, I feel, than using petrol. When the bees are dead you then break up the hive into sections and burn them. Although the hive is thoroughly destroyed, the virus can survive for another 50 years which is why the pit should be deep enough not to be disturbed if the ground is subsequently cultivated. (Even a plough will not penetrate this far down.) Burning the hive is acceptable if you use wooden hives, but what of polystyrene? It is illegal to burn something as noxious as polystyrene. What you must do is to destroy your bees and wooden frames by burning them, but then take the remaining contaminated parts and totally immerse them in a cleaning fluid like Milton or, better still, a solution of caustic soda, diluted sufficiently so that the soda burns a thin layer off the hive and kills the virus.

'Ropiness test' with a match stick

So much for dealing with AFB, but how do you get it in the first place? This disease is caused by a spore-forming bacterium called Paenibacillus larvae. It's a nasty piece of work. Bee larvae which are no more than three days old are infected by ingesting the bacterium spores that are in their food. What happens then is that spores germinate within the gut of the bee larvae and the vegetative form of the bacteria develops, essentially eating the larvae to survive. Once the spores have taken hold it's curtains for the larvae. The infected larvae will darken and die soon after. Larvae more than three days old are safe and secure but such is the strength of the spores that infection can spread quickly and from hive to hive. Although the vegetative form of the bacteria will die after three days, the spores can remain viable for many years. A dead larva can carry as many as 100 million spores so even if the hive may appear clear,

there is a time-bomb waiting to explode.

Can you see why throughout this book I have been harping on about being vigilant for disease? It's time for more tea I think; it clears the head. Everything looks better with a cup of tea in hand. Sweetened with honey, of course.

What must I look for and how will I know if my hives are infected? Well, testing in a laboratory is the key and your bee inspector will deal with this and instruct you thereafter. Often you only realise you have AFB after a routine check by your bee inspector, but if you suspect that you have AFB there are tests that you can do 'in the field'. Let me give you the characteristics of American Foul Brood contamination. The sealed brood comb will be sunken, discoloured or will have punctured caps. You can usually see a dip in the cap or sometimes a tiny pin prick - usually only one. (Once, when I was a novice bee keeper, I thought it was the opposite, only to find a new bee was emerging through the tiny hole.) The colour of the dead brood is usually a dull white to a light coffee-brown. The colour varies, however, and they could be a darker brown or even black. What they don't look like is a healthy young bee. Their bodies or mass will be soft, unformed with a slight stickiness, and they may, in later stages of contamination, appear to be stringy. What AFB does have, which is a real indicator, is an odour, which smells putrid. You may wish at this stage of your field trial to take a pin or match (or if you don't have any of these to hand, use some straw or a thick blade of grass, or better still, a discarded bird's feather). Dip the end of the pin or match into the cell, breaking the cap and run it (the implement) to the bottom of the cell. You will find a sticky, almost resin-like substance. A strand of brownish, smelly, liquefied larvae will be drawn out of the cell. Another thing to look for is scaling. Usually scales form on the inside of the contaminated cell and they lie uniformly flat on the lower side of the cell wall. Both of these are a sure sign of AFB. Here we are visiting another planet, and it's certainly not Planet Craig. We will call this one Planet I-Want-To-Get-Off-Now-Please.

The disease spreads easily, hence the strict legislation controlling the movement of hives. One way that the disease spreads is through the bees themselves. This can happen in several ways: firstly, by cleaning infected cells ready for further use, the bees unknowingly distribute the AFB spores throughout the colony. Secondly, as the bees attempt to remove the infected larvae they contaminate the brood food and any nectar that is stored in the contaminated cells will contain the dreaded spores. Thirdly, as the bees move the honey up into the supers the

entire hive becomes infected. Eventually the hive will become weak and the robbers will soon appear. This is a disaster because the wasps, bumble bees and other honey bees that are stealing the contaminated honey will spread the infection and contaminate their own stores and young.

Imported bulk honey can also be a source of contamination. If the containers in which honey is transported are not washed out properly, any honey residue can be taken by robber bees who will then take the honey back to their own hives and the disease will become established.

Another way that the disease can be distributed is by the bee keeper him/herself. As you move from hive to hive you may unwittingly spread the spores. It is therefore terribly important that when you visit another hive, even if it is one of your own, you sterilise your equipment in between. The easiest way is to place your hive tools, brushes and so on, in a solution of sterilising fluid. (You can also sterilise your hive tool over a gas burner, but obviously this doesn't work with other tools.) If you come into contact with an infected hive, change and wash/sterilise your whole kit - bee suit, gloves, everything. I cannot stress too much just how important total cleanliness is. Trevor is here so it must be tea time.

There is a theory that AFB is present in every hive and that the bees work alongside it successfully, and it's only when the brood becomes weakened and stressed that the disease takes hold. This is where the theory of moving bees too far and too often, and leaving them on a mono-diet comes into play. Imagine you are fed the same food, day in, day out and are moved through time and weather zones on a regular basis: you are highly likely to become stressed, upset and liable to become prey to diseases. It's the same for bees. It is far better to give them a varied diet and keep them in one place for as long as possible.

You can treat your bees with antibiotics and in the USA, the Food and Drug Administration has been successful with the likes of oxytetracycline hydrochloride, or tylosin tartrate. In China, the vast use of antibiotics for many years appeared to cure the spread of AFB spores, but in fact it was only masking it. The UK bee inspectorate, however, prefer a policy of burning the hives to eradicate the spread of spores.

European Foul Brood

European Foul Brood (EFB) is, in some ways, similar to, but it is generally held to be not as bad as AFB. It is a bacteria that infests the gut region of an infected larvae and, like AFB, is extremely detrimental to the hive's growth and the colony's strength. The difference between the two is that EFB is often considered to be a stress related disease (can you see a pattern developing here?) and is not spore forming like AFB. When on the lookout for EFB, look for dead or even dying larvae, which on first inspection appear to be curled upwards. They may be brown, but not a dark, chocolatey brown; it is more of a light,

coffee brown or even a light yellow. You will find that the larvae may be deflated, or even liquefied as if they have melted. Some may have their tracheal tubes pressed forward. You may even find some of them dried out and with the texture of rubber.

I said earlier that this is classed as a stress-related disease and is only dangerous if the colony is already under attack from stress for other reasons, like too much movement, a poor diet, too great a distance to foraging sites, robbing and so on. Quite often a strong hive can survive and live quite easily alongside EFB. An outbreak can be controlled with the use of oxytetracycline hydrochloride. Although you will have to be careful with any of the stored honey as those chemicals for treatment may be present within it. An alternative to using chemicals is to shake your bees off the comb and replace your old frames with new ones. A belt and braces approach would be to introduce new frames and replace the hive. You can then sterilise the old hive with a blow torch or sterilising fluid.

European Foul Brood

FUNGAL, VIRAL AND OTHER DISEASES

Chalk Brood / Stone Brood

There are lots of fungal diseases to be found in the hive. This is because it's a nice warm moist environment for bees and uninvited guests to thrive. You might think that you had reached the apocalypse with AFB but this little blighter will compete for food and nourishment with the larvae, and ultimately the fungus tends to win, causing starvation for the baby bee. (I like to describe them as baby bees and not as grubs or larvae as I feel it makes the bee keeper appear/feel paternal/maternal and responsible for their babies. I know mine all by name.) Once the fungus has killed the baby bee it will continue to 'eat' it and eventually what's left of the baby will appear white and chalky – hence the name 'chalk brood'. As for stone brood, it is, as its name suggests, similar to its relative chalk brood. Again it's a fungal disease that causes

Tips!

- *A match stick is all that's required to do a basic assesment for American Foul brood.*

- *Larvae spores/fungi can be dangerous to the respiratory systems of animals and humans alike.*

- *Try some honey in your tea!*

the larvae to die and become mummified. When the poor unsuspecting larvae take in the spores, they generally hatch in the gut. They grow rapidly and form a collar near the head of the baby bee. They die, turn black and become difficult to crush - hence the name. In the end the fungi erupts from the larvae and forms a false second skin.

It's at this stage that the larvae are covered with a powdery white dusting which are the spores. The fungi like to inhabit the soil and are also found in mammals, birds and other insects, so they have a wide range of hosts. The fungi aren't easy to identify at the early stage; although the spores of different species have different colours

Bee larvae/Craig's babies!
(Right cell bees older than left cell)

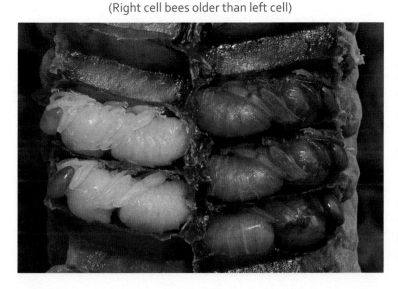

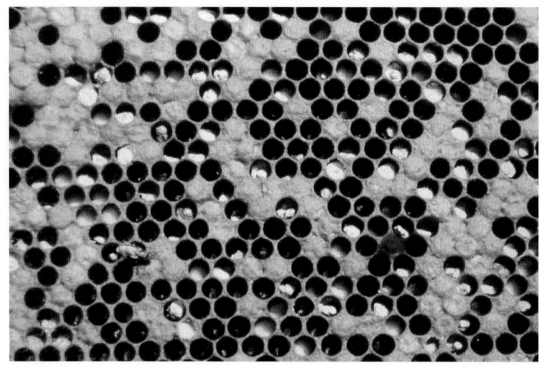

Chalkbrood

they are not easy to tell apart as they hold no common grouping. Beware that these spores can be dangerous to the respiratory systems of both animals and humans alike.

Worker bees will clean out the infected brood and depending on the level of infection the colony may recover, but this will depend on a number of factors, one of them being the trait of the individual bee. Some bees are more hygienic and house-proud than others and this will have a direct bearing on the cleanliness of the hive. The cleaner the hive, the less likely there is to be disease.

That seems to take care of our chums the fungal diseases, but is there anything else that you should be worried about? Unfortunately yes. I want you to just take a little bit of information on board here please. I am telling you about the problems with keeping bees so if you are midway through the book and are thinking about packing it all in and keeping a pig instead, well I can sympathise with you. But if you do keep any other livestock you will need to be with them all the time, being a vet, looking after them, looking at their soil content, examining their mess. And pigs can maul. But my point is you have to be aware of everything that can affect your 'pets' even if you don't encounter them directly. So onwards and upwards – with another cup of tea.

Viral Diseases

I have decided to include these as I feel you may come across them in conversation or on the internet and you may like to know a little about them, even though many of these are not found in the UK and/or in the USA. **Acute Bee Paralysis Virus (ABPV)** is a common infective agent. It is related to those other infective agents that cause other viral diseases such as **Israeli Acute Paralysis Virus (IAPV), Kashmiri Virus (KV) and Black Queen Cell Virus (BQCV)**. You don't need to worry about these as they are not found in the UK, but it is as well to be aware of them. You can find it in a healthy colony and the bees appear to live alongside it fairly well. Please don't think that this paralysis is a major concern but it can be worrying when you open a colony to find they are all a bit sleepy or not moving. It is believed that the cause of the paralysis may be down to the excessive amount of pollen and nectar from the likes of buttercup, laurel and rhododendron. It's interesting to note the effects rhododendron pollen and nectar can have on bees and humans alike. In the eastern part of Turkey there are a range of mountains (I'm not going to say which ones), which are full of rhododendron bushes. It's a beautiful cacophony of colour and smell. When the snows melt the bee keepers take full advantage of this blossom and move their hives up and down the mountains to allow their bees to forage on nothing but rhododendron. Innocent you might say. Well as they start off on the lower slopes they find the honey made there is very good; nice colour, flavour and consistency - a good all-rounder. As the weather improves the bees are moved up the mountainside to a point at which the temperature is just right for bees to forage. This honey is poisonous to humans but the honey produced is stored and used as a winter food for the bees and therefore the bee keepers don't have to buy expensive fondants or syrups. As the weather changes and the mountain tops are exposed to bees, the bee keepers move their hives up. Now you are probably thinking, will this honey be poisonous or edible. Well it's edible, with a twist. This type of rhododendron honey

has the same hallucinogenic effect as LSD. Truly. I was once in Turkey and asked a shop keeper about this honey. Ah yes, he said, I can get you some. Days later he produced a dark, almost sweet chestnut type honey. As I went to try it he told me not to as he would get into real trouble. I took it home to England and set about warming it up for a little taste. I certainly don't take drugs but I was fascinated that a honey could have this effect. It didn't, so I took a larger spoonful: nothing. In the end I finished the honey and nothing had happened. I was told later by another older and wiser bee keeper that this happens a lot in Turkey and I had been sold Sweet Chestnut honey. The honey from the tops of the mountains is (apparently) hallucinogenic, but the Turkish government have banned the sale of it.

Conversely, the virus may be caused through a deficiency in pollen through brood rearing during early spring ,or more importantly the consumption of fermented pollen. Bees affected by ABPV shiver like there is no heating on. They tremble and can't fly, lose hair from their bodies and have a greasy, dark, almost shiny appearance. We are talking about a lot of bees doing this here, not a couple. Bees in this condition are prone to attack, are submissive and can be found (when it's serious) crawling up the sides of the hive, on blades of grass and falling to the ground as if they are drunk. Israeli Acute Paralysis Virus has been associated with **Colony Collapse Disorder (CCD)** which I will talk about later. There are a number of other viruses worth mentioning but not worth exploring too much. Kashmiri Virus has recently been found in Denmark and is related to IAPV. BQCV, CPV, CWV and **Deformed Wing Virus.** All follow a similar pattern in that it is believed that the virus is transmitted through contaminated foods.

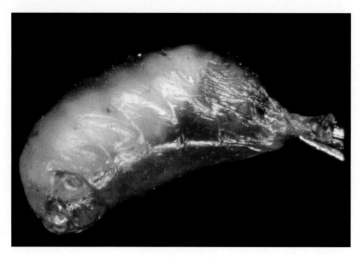

Sac brood

drops too quickly to a critical point the nurse bees will be unable to protect the babies and you may lose the colony altogether. Choose your timing wisely.

Pesticides

Sometimes bees are lost through the use of pesticides. This is a problem both in the countryside and in urban environments. I generally know what the farmers are growing and therefore what chemicals they will be using. But the urban bee keeper generally doesn't know what Mrs Smith next door is using. Some carbamate pesticides can be especially dangerous as they can take up to two days to show up on the 'problem register', by which time the 'problems' have resulted in a decline in bee numbers. (See www.edis.ifas.ufl.edu/pio88 for more information.) The likes of organophosphates will kill bees quickly. In many ways you will be troubled to find out about these as they used to be used a lot in the countryside, more so than in the urban areas, in fact. I don't know what the situation is regarding organophosphates in America, but those of you on the edge of the towns may find your bees come into contact with organophosphates in the open areas. What you will find when pesticides are present in the hive is a sudden and drastic loss of bees in the colony. You will usually find the bees in mounds in the front of the hive. You may also find dead bees on the ground corresponding to the flight path as they have been unable to reach the hive and have died or are dying. From a positive view, if they have not been able to get back to the hive and have died

Another virus - **Sac brood** - changes the colour of the bees from a white to a grey and then finally black. When the affected larvae are removed from their cells by the housekeeping bees, the cells appears to be filled with water, hence the name. Let's have a break from nasties with a cup of tea. A delicate darjeeling I think to wake me up a little.

Chill Brood

Fortunately it's not actually a disease but more to do with the unintentional mistreatment of the bees by the bee keeper. Chill brood can be caused by a rapid drop in temperature. You will know the scenario: you see the sun, it's nice and warm and you start wearing t-shirts. Then overnight the temperature drops and it's snowing. That sudden drop is a killer. The brood needs to be kept warm at all times so you will find the nurse bees clustering over the larvae trying to keep them warm and at the correct temperature. I know it's tempting on a warm, early spring day to look into your hive to ensure every one is having a good time, but if the temperature

outside, at least they haven't brought any more toxins back to the hive.

Colony Collapse Disorder (CCD)

Well what can I say? This is probably the best known disease/disorder/issue surrounding the plight of bees. It is a subject that so many people seem to know about and are prepared to give advice on. Let's look at the facts and try not to be too worried. CCD was initially found in bee hives of the western honey bee and mainly in the northern hemisphere. Initially this problem was found in North America, where hundreds of hives have been lost. Bee keepers would go one day to look at their hives and they would be healthy, full and brimming. They would return the next day and the majority of the bees would be gone - and where to no-one knew, almost like elephants to an unknown graveyard. Various agencies and governments had an opinion. Even here on Planet Craig I had my own thoughts. In Europe, similar problems were starting; in Portugal and north as far as Holland and as far south as Greece. Some cases were also found in the Far East. The cause is so far still up for discussion. We have heard a thousand and one reasons as to its source. I have heard everything from the effects of mobile phones transmitters and electric cable pylons, malnutrition, GM crop forage, Israeli Acute Paralysis Virus and others such as the disease and problems associated with the varroa mite. In a recent publication from the Science Daily there is a claim to have possibly found a cure for the problem. It is believed that a bacteria, Nosema ceranae, has been causing the trouble. Scientists looked at two professional apiaries in Spain that had apparently suffered

from CCD. They looked at the health of the hives and noted that apart from some issues with varroa, there were no other diseases present. Initially IAPV was considered but this was discounted. The bees were treated with an antibiotic, flumagillin. It appeared to work a treat. The colony appeared to recover, but I think more research may have to be done before we can say that this is the answer.

My thoughts are that, (and this is a theme running through the book), bees don't like being moved. Commercial bee keepers who trek their livestock mile after mile aren't doing their bees any good. Some commercial bee keepers in America will move their bees thousands of miles from the Atlantic to the Pacific and will make their bees work twelve months a year. I have fairly good success with my bees, but I am a small commercial bee keeper and I only like to move my bees a short distance, maybe twenty miles in total. I also think there are problems with mono-diets. Bees, like any other sentient beings, need a balanced diet, so to restrict them to the same forage for weeks on end is not a healthy option. As urban bee keepers you should not have a problem providing your bees with a varied diet. Place your hive sensibly and you should be fine.

CHAPTER THREE
SWARM CONTROL

This is probably the greatest fear of a number of bee keepers and certainly of the general public. They (the public) will hold an almost Hitchcockesque fear of a swarm. Yet when bees swarm they are probably at their quietest and are very docile. In the northern hemisphere swarming usually takes place between late April and early July. If you have one hive and it has a yearling queen in it, she won't be laying any baby queens. So your first year is free of the worry of swarming. In her second year she will want to lay loads of daughters (queens), in her third year, some more but not as many, and hardly any in her fourth term, if she is in fact still alive. Swarming is a perfectly natural occurrence, but when you are dealing with tens of thousands of bees, it can be a little disconcerting. When bees swarm, it's like cell division - twins in the womb, bacteria dividing — except that the division is on a greater scale and the bees will want to leave when there is no more living space available. It's merely nature reproducing on a large scale and a new colony is formed when the queen leaves, accompanied by a large group

Supersedure Cell

of workers. In the initial swarm around 50 to 60 percent of the bees will leave with the queen. Oddly, you might get an 'after-swarm'. This is a little like an earthquake with the main shock and then a smaller after-shock. I have experienced this in the past, but it is not a common occurrence. Perhaps I shouldn't have described a swarm in the same way as an earthquake. If there is an 'after-swarm' it will be smaller and will contain one or more virgin queens (queens that have not yet mated).

How does all this start? Well the workers prepare queen cells throughout the year and when the hive is ready to swarm, generally through overcrowding, the queen herself will lay queen eggs in these specially prepared cells. Queen cells hang or protrude in a noticeable fashion. The egg cell is larger and looks like a monkey nut. Worker cells are the smallest and drone cells are slightly larger than the worker and both fit the frame, unlike the queen cell. As a new queen is raised you should be aware that the hive may swarm immediately she emerges, or she might take stock for a couple of days. Generally the queen is too heavy to fly any further than a few yards. In the meantime the workers will stop feeding her and send her to the gym to get ready for the hive to split. The queen gets into shape and stops laying eggs. If you have the opportunity to watch your hive you might notice scouts moving backwards and forwards trying to identify a suitable place to swarm to. It can be anywhere, but

is often in the most awkward place. You generally find that although the scouts look for a new home, the queen and her swarm will land on any given place, which is usually accepted by the scouts and there they will stay. Sometimes as a bee keeper, you can look at a settled swarm and think, yes, if I were a bee I would settle there. Other times you just shake your head in disbelief. I once knew a bee keeper, an elderly chap, who couldn't resist the opportunity of a new colony. He was called to a swarm one May afternoon; it was up a tree, close to a school. The Police cordoned off the road and up went Arthur. As Arthur neared the bees he smoked them, positioned his cardboard box, clutched his heart and fell 60 feet out of the tree. Everyone thought he had been stung, but in fact he had had a heart attack. I can't decide whether it was the fall that killed him or the heart attack, but either way he died enjoying what he did.

I was once called to a swarm in a post box. Clearly I couldn't get my hand in and the postman didn't want to come anywhere near it. I asked him for the keys and explained I would get the bees out. He then told me as the keys were the property of the post office and therefore the Queen, I wasn't allowed to have them, so I had to await a supervisor from the post office depot for him to OK the opening of Her Majesty's post box by a member of the general public. Great fun as nature doesn't have any sense of tradition or etiquette.

A point to note: if you have hair, try to keep it a little longer on top - it gives you that three second bonus for when the bees decide to attack from above and you don't see them. If you are challenged in the hair department, wear a cap.

What makes the queen and her entourage choose a new site? Well, when the swarm decides it's time to go, they won't just head off into the blue yonder, they will hover around the hive for a while. They may then appear to fly, but return backwards and forwards, as if they have not quite made up their minds. When they have decided it's time to be off, they will head for the nearest semi-suitable place and a small cluster will form around the queen, keeping her warm and making sure she has a nice cup of tea and a slice of battenburg. She sends off a number of scout bees, anything up to 50 or 60, to survey the area for suitable sites. The scout bees are chosen because they are the best foragers in the colony. Eventually a single scout will return and will perform a dance similar to the waggle dance you might have heard of. This will indicate the distance and direction of the possible new site to the rest of the new colony. A point to note here is that the more excited the scout is about her findings, the more excited her dance. If she is successful in her dance and can convince her fellow workers through her enthusiasm, then there is a good chance the rest of the bees clustered will follow. Initially the other scouts will join her to look at the property with a view to buying. When all are happy, off they will fly. This maybe for a couple of miles or even further and it is a sight to be seen - as long as they aren't your bees! They take shape at the agreed site, start to build a new nest and secure the colony as at this time it's a fairly vulnerable position to be in if you are a bee. When you think about it these scouts not only have to find a place that is potentially suitable, it has to be secure, waterproof, at a good angle to the sun, free from ants and large enough for expansion. So ONE scouting bee has to find somewhere suitable and will also have to make an assessment of the possible future

advantages for the colony, bearing in mind that the bee making this decision will be dead in a couple of weeks. A swarm that is on the wing will only have sufficient honey in their stomachs for the energy they need to fly and to start the colony, and sometimes this is not enough. It is feasible that a swarm will starve or easily die if the weather changes considerably. I have come across a swarm that has left the colony on a nice warm day and the weather has changed for the worse. The bees have slowed down to the extent that they are unable to defend what little structure they have, they are unable to send out workers to forage for any nectar and the honey they have stored in their stomachs is being used up to create warmth to keep the queen alive. They have a tendency not to leave her and concentrate on keeping her warm: if the sun isn't able to assist, they are in real trouble. This can be a problem if you are keeping bees on a coast with a prevailing wind. The west coast of the United Kingdom can be tricky with the prevailing westerlies.

Tips!

- *A new queen's first year is likely to be swarm free as she is unlikely to make any queen cells.*

- *Ensure you add brood boxes and 'supers' to your hive in the queen's first year to ensure good honey flow and decrease the likelihood of overcrowding.*

- *If you only have one hive and no space for a 'back up' hive, you may have to contemplate culling any queen cells that you can find in order to prevent swarming.*

the only change in their lives is that they have shrunk by about half. The new queen that is left in the original hive will need to fly and mate on the wing, unless they are artificially inseminated. This in turn provides a number of problems. At times on her mating flight she can get lost and die, can be eaten by a bird, or killed by a sudden cold blast of air. This is disastrous for the remaining bees as there will no longer be any young brood coming through, and with no young brood to raise any queens, the hive will struggle, robbing will occur as the colony shrinks and eventually it will die.

How do I manage the bees at swarming time or are they managing me? A bit of both really. You can essentially guarantee that in the new queen's first year she won't make any queen cells, so your first year of bee keeping will be swarm-free and hopefully a little easier to manage and free from worry. Ordinarily she won't lay any queens, but she can swarm if you don't follow the rules. As

Before today I have found swarms and lifted them with my bare hands, popped them into a cardboard box, sealed the top and made some air holes for them to breathe and then brought them indoors and placed next to the Aga or a radiator to allow them to warm up. I put a little honey or some sugar and water on a saucer for them to eat. You will know when they are ready to leave home as the noise from inside the box will change and they might sound a little cheesed off. This is why it's important even if you only have a single hive to have a spare one available. Leave the spare hive a few feet away from the potential swarmer and place some swarm attractant in the hive, full of nice new frames and marked with this year's colour. You can buy the attractant from most bee suppliers and nature, coupled with a little help, should do the rest. Those bees left behind in the original hive are fairly secure;

If you don't want swarming to take place and you are not interested necessarily in having several other hives, go around your current hives every ten days or so and inspect them for queen cells. If you find any you should just squash them. Be vigilant - have a look all over the hive, including the side walls and under the roof as they can lay anywhere and it will always be the one place you have missed where there will be a queen cell that causes the problems. Personally, I think this is a bit of a waste of time because undoubtedly there will be one that will go undetected and you will have the swarm you didn't want. For me, from a commercial basis, swarms and a good queen-laying queen are perfect, but if you have one hive and wish to keep it that way then it's understandable that you will want to cull your queens.

the colony is growing, unless you are adding brood boxes or supers for the honey flow, the queen may be tempted into laying a queen and will then decide to find somewhere a little less crowded. In year two she will be at her prime and will be laying queens at a rate of knots. Certain bees seem to want to swarm more than others: German carniolans have little interest in swarming, Slovak carniolans a little more so. Yet one year my Scottish Blacks appeared to be swarming for fun and I could not keep up with them.

If you do cull, remember you may have prevented an immediate swarm, but in the majority of cases the queen will merely start again and build more cells. You could repeat the process again and again, but that defeats the object. As much as we want to think we know and understand bees, most of the time I think that they are observing me and not the other way around. We can not really predict the way that the bee is going to react. If I said that the colony would definitely not swarm if you carried out a culling exercise I would be lying, but then on the other hand, they may not swarm. We just don't know what makes the colony decide that it wants to swarm - if I did I would be incredibly rich as well as drop-dead gorgeous (well, at least the former, anyway!).

In many ways all that you can do is buy yourself some time. Swarming is a natural process and although I've never been happy playing God, I have used a method to prevent swarming. This is a process that involves snipping the queen's wing. This should be done with a pair of sharp, rounded-ended nail scissors. Don't use large ones as it's a delicate job and you need to be in control, and don't use ones with a point because you will either stab the queen or cut one of your fingers off. I said wings, plural, but in fact you just cut one of her wings diagonally. Clipping the wings of the old queen will physically prevent her from flying. Imagine if you shorten one wing of an aircraft or glider; when it is airborne it will spiral downwards. This is what will happen to the old queen. The clipping of wings will not upset or injure the queen at all so when she is ready to swarm she will leave the joys and warmth of the colony but will only be able to fly a short distance. And I'm talking feet

here. Because she can't fly too far her main entourage won't follow her and you will be able to watch them cluster around the hive like people waiting for a bus to arrive. She may try several times to jump ship, but she will still only travel a short distance. She may die of exhaustion in the grass or she may land on part of the hive and the workers will mob her to form a ball to keep her safe and warm and to encourage her to swarm. But it's useless. The bees that have gone before her will soon return to the hive and start the process without the old queen. It will be a little time before the bees can look to swarm again as they will have to wait for one of the virgin queens to emerge before they can leave. It is because of this wing clipping process that the 10 day rule is applied. Let me explain further.

The Ten Day Rule

The ten day rule works like this. When you are on an inspection of your hive check on the status of the queen cells. If they are occupied by little queens you have to work on the basis that the colony has plans to swarm. If the queen cells are not sealed but the eggs are in situ then you can work on

the basis that the colony has not swarmed yet. You can be almost 100% sure of this as bees will usually wait until there are sealed cells before they decide to leave. If you are happy with the queen you have and you are not necessarily looking for maximum egg laying and don't want to re-queen too often, you might want to find the existing queen and squash all the remaining queen cells. This

does feel somewhat negative as there are so many people looking for queens and you are dealing with a species that is struggling to survive. But you must decide. Remember this "squashing" exercise is only going to buy you time and will not solve the problem, because after the queen has realised her daughters are not hatching, she may instruct the workers to reshape either the old damaged queen cells or make new ones. What we have to work on is that although we have sent men to the moon, invented the internet and learnt how to conquer flight, we know next to nothing about all bees. Having removed the queen cells you may find that the bees remake the queen cells and give the impression of looking to swarm, but then the queen chooses not to lay any eggs or the workers decide to tear down the cells and use the wax elsewhere. You just can't tell. My point is that we can only guess at what they are thinking and can only minimally alter the course of their actions. I put it down to the fact that the majority of the bees are women.

If the colony is going to swarm and the queen has replaced the queen cells with new, then we do know that the bees won't look to swarm until the cells are sealed. How long is this then? Well if we have just removed the queen cells and we believe that the queen is going to restart laying in the queen cells then we can calculate from this point. We have three days from laying until the eggs hatch. From then on the larvae will be fed for six days. Working on this basis it is at least nine days before the queen cells will be packed with special queen food and sealed. So is this what I mean about the ten day rule? Not entirely. All larvae are fed the same food for the first three days; they all have a taste of the royal jelly but only the baby queens manage to get any more after that date. After day three the workers go on to the equivalent of porridge and gruel whilst it's caviar and champagne for the queens.

What you need to know at this stage is that if the worker bees choose a larva which is no older than three days old, they can change her status from lowly worker to queen merely by changing her diet. Older than three days and it's too late, and the bees are destined for a life of foraging.

If the bees do this straight after you have gone on your squashing spree, the bees can have sealed cells in the colony within three days. On this basis the queen will go and you will have lost half your hive. The answer is to work through your hive every three days and check that there are no more queen cells and if there are, kill them. Doing this over and over again is time consuming and laborious though, and won't prevent the colony from swarming. My advice is to go back to the start of this section and clip the wings of the queen.

A number of events might unfold here. The clipped queen might leave, fly around and die in the grass or get blown away on a breeze if you are lucky. She may make it back into the hive and try over and over again to get her colony to leave. Or she might be able to make her way back to the underside of the hive and a ball or cluster of bees will form around her to keep her warm and secure. One thing we do know is that the workers can't leave the hive with the old queen still in it. These workers will have to await a virgin to emerge and this won't happen for ten days, hence the term. Now you have bought yourself ten days breathing space. If queen cells are found you might want to try an artificial swarm which is covered next.

Another method of preventing swarming requires the removal of capped brood along with the old queen; you will know it's the old queen because of her markings. The frame is placed into a hive with empty but drawn frames on it. The old frame complete with brood and queen is placed in the same location as it was in the old hive. A super is placed on top of the hive box and then covered with a crown board. You will then need to inspect the old hive for any new young queens and if there are any they must be destroyed. The hive box with the bees in it is now placed on top of the crown board. Any bees already out foraging will return to the lower box and will by sheer numbers deplete the numbers in the top box. Leave it for around a week and then inspect it again. Any further queen cells should be destroyed. Leave it for another week and any desire by the queen to swarm is usually removed as the other queen cells are destroyed and are no longer any threat to the old queen's existence. You can now reunite the hive completely and normal service should be resumed.

The Artificial Swarm

My Uncle Nipper was a big, heavily built man from near Wigan. And yet he was as gentle with his bees as he was with us as children. When I was about twelve he told me about the Artificial Swarm. He talked about it in such a way that I felt it should have capital letters, and that I had been initiated into the bee keepers' secret society. It felt so technical and adult. And he was trusting me with this secret. The truth was I could have found out about this in any book on bee keeping from the library but my Uncle Nipper imbued it with all the mysteries of a Tolkein tale wrapped up with Harry Potter and his friends. My Uncle told me that if you find yourself with a colony building up very fast and despite all your previous efforts to prevent or at best control swarming, you feel that swarming is inevitable, you may be inclined to go the 'artificial route'. This process allows you the opportunity to split

Tips!

- *Clipping the queen bee's wing will help prevent swarming and causes little/no distress to the bee.*

- *To clip the wing (just one), use sharp, round ended nail scissors.*

- *The wing should be clipped on a diagonal angle.*

the bees without them actually swarming. Because you, the bee keeper, are making the decisions as to what shall go and what shall stay, you must keep an eye on what is needed for both the old and new colony. Uncle Nipper told me that he only ever split the hive in two, but he also said that you could divide and subdivide depending on the strength, age, security and stores the bees have. I have always followed his route and stuck to a simple division. If you do decide to carry out this process you are better to work on the basis that this is what you are going to do and not to be distracted away from your mission. It's now when your spare hive for swarms or emergencies comes in handy. So remove the roof of the original hive, set aside and remove the super and queen excluder if there is one present. Then remove the brood frames and check them thoroughly for queen cells. When you find the unsealed queen cells with eggs and larvae in, have a look for any 'ordinary' eggs on the comb. If you don't find any eggs, what does that tell you? Well it means that the queen hasn't been laying for at least three days and in that instance I would be looking to see if the queen was still

active and present. Find the queen and see what state she is in. When you find her she must be placed to one side, preferably in a queen cage with a couple of attendant bees which will allow the bees to identify her but will not frustrate or upset them too much. When you have done this it's a question of fiddling around with the frames. First of all, however, take the original hive and place it somewhere else in the apiary - its location is unimportant at this stage. And where the original one was, place the new, spare one, complete with a full set of frames. It should be positioned in exactly the same place as the original one, so before you move the original, mark precisely where it was - even the slightest variation can be detected by the bees. You must check and, take my advice, re-check for any queen cells still on frames being returned to the original brood box. Equally, whilst the frames are out have a look around the brood box for rogue cells. This frame is then returned to the centre of the original brood box.

At this stage my uncle would take the queen from her cage and place her directly on to a frame to be placed in your spare hive. If you have the luxury of drawn comb then place at least three of them on either side of the frame of brood with the old queen on it. Fill the remainder of the brood box with comb containing empty cells and stores. You can just use foundation, but it's not ideal, and I wouldn't unless you really have no frames with empty cells. If you need to, take a frame or two of food from the original hive. The new brood box with the old queen in it will have plenty of space for new brood and some food in case of poor flow. Now my uncle would always place the queen excluder and the supers from the original colony in the new hive. The queen is now in the same position as if she had swarmed, although she has the added advantage of having drawn out comb, some stores to survive if the weather takes a turn for the worse and a ready-made 'nest'. If the queen has slowed down in her laying whilst getting ready to swarm, she will soon be back to speed.

At twelve this 'artificial' swarm was my reward for helping Uncle Nipper - my very own hive. But now back to the original hive. Have a look for a nice big fat juicy queen cell in the brood box. How do I choose one, I hear you cry. A good question. Well you should look for large cells in the middle of the comb with lots of nice worker brood surrounding it. The middle is generally best as it maintains a constant temperature all round, whereas those on the periphery can become cold and be susceptible to a fluctuation in temperature which overall isn't good for a queen cell. You may find that those on the edge are somewhat weaker in nature. If you find a cell is not sealed, look for one with lots of royal jelly in it and a big grub. Avoid, if you can, very long cells or short, small ones. Also try to avoid cells with a smooth finish to the cell - this sometimes indicates that very little attention has been paid in the construction of the cell by the workers. A cell with texture and rippling indicates that it has been constructed more carefully. Once you have decided on which queen cell you want, place the frame with the queen cell to one side and destroy all the other queen cells. Now give the frames a quick shake to dislodge the bees on the frame and whilst you do this make sure you are directly over the new brood box. You can now reassemble the hive.

Always make sure there is plenty of food. This can either be in the shape of store food or you might wish to use an 'artificial' food which I talked about earlier in the book. If you use the latter, keep an eye on the weather. If a good spell is due when the bees can forage to their hearts content and you are not going to be hindered by a frost, remove the feeder and let them work for themselves.

Within the ten day period check to see if your queen has vacated the queen cell. If she hasn't, she may have died or been killed, in which case you will have to start all over again. Take comfort, though; this isn't common so don't worry too much. If this did happen and you didn't want to start the process all over again, buy a queen from a reputable dealer. Leave her in the cage for 24 hours so that the resident bees are aware of her and her scent, then open the base plate or hatch on the queen box. Within three days the bees will have become used to her scent whilst the scent of the old queen will have gone. Eventually she will mate and off we go. Be patient.

It is as well to keep your records up to date at this stage. Now is the time to mark the queen and record her age. If you don't want to physically mark the queen you can mark the hive instead by placing a simple coloured thumb tack on the top of the hive roof. At a glance you can see what year the queen is from by using this mnemonic:

Be Warned You Require Gloves!

| Blue 2010 | White 2011 | Yellow 2012 | Red 2008 | Green 2009 |

Place a marker for last year's old queen on the original hive and this year's colour on the new hive. If you want to start expanding your colonies this system will help tremendously and save an awful lot of wasted time and effort. A yearling will not be laying any queens, so you can leave these hives alone. A second year will be laying at her maximum and will be at peak performance; by year three she will be slowing down and if she is still alive by year four, she will be laying at a very slow rate. You might want to re-queen every second year so that your hives are strong and active.

When it comes to capturing a swarm there are a number of different methods that bee keepers swear are the best. You can use a Nasonov pheromone to attract bees to a hive, which is quite effective, or, if they have only recently swarmed, you can collect them easily by hand. Many people will talk about using a swarm box or empty hive in which to collect a swarm, but if you are up a tree and on your own it's quite difficult to carry around everything you need. My advice is to use a cardboard box. One that is robust, dry and not too big. Too many accidents are caused by having to go up a tree or by digging underground to try to locate a swarm and you either have too much kit or not enough. Mmm - tea. A nice bit of Gunpowder green, I think.

A swarm up a tree can be tricky, but I swear that bees will choose the trickiest place they can find. Get up the tree with your box, smoker (preferably lit) and something like a shopping basket or a hessian sack to put the box in. Have a length of rope attached to the basket or sack so that you can lower it down gently. Have a roll of tape to hand - gaffer or duck is good. Get to where you need to be up the tree and edge as close as you can without worrying the bees. Give them a good smoke and place your box underneath them. Give the bough a good slap and allow the bulk of the swarm to fall into the box. Close the box, lower it down and then climb

down. If you have an audience, smile just to show them that it was dead easy and you are in fact a super hero and gratefully take any kisses from young ladies nearby. Some bees will be flying around but it doesn't matter about these; you only need to worry if the queen isn't there, in which case you need to try again.

There is another way of catching your swarm and I do like to use this method. It makes you look like the James Bond of bee keeping. Place a white sheet on the floor directly under the swarm with a hive box and floor to one side. Confront your swarm, not with a smoker but with the opposite - a spray gun filled with a warm, sugary solution. Give the bees a good blast on all sides, allow it to cool and then give the branch a good slap. The solution should hold the bees together, give them something to eat in the short term and keep their minds focused on the immediate food. As they hit the ground in a cluster it should break. The glare from the sun on the white sheet should make the bees run for cover and head inside the hive, where they can avoid the strong sunlight. Here in the dark the bees are happy and will quickly adopt this as their new home. To assist the bees in going into the hive you can put a short wooden plank with a sugary solution on it propped up to the entrance. The bees will be attracted to the sugar and the ease of entering the hive and as we said earlier, Roberto et to azia. Bob's your uncle.

You can also use a suction pump to collect swarms. The Royal Parks and Gardens in London use these all the time because the Health and Safety Executive won't allow them up a ladder, so it's all done with what is in effect a big vacuum cleaner. If you are thinking of using a vacuum cleaner, don't. I did once and my wife nearly killed me.

CHAPTER FOUR
QUEEN CELLS

Let me just explain a little about queen cells. A bee keeper asked me many years ago about the contents of a hive I owned. I said I was just going to kill the new queens off before they hatched and maintain the status quo. Hang fire, he said - there are 'queen cells' and 'queen cells'. I was just going to squash them and get on with managing the hives. But it wasn't as simple as I first thought. He explained several times to me that there are three different queen cells. He continued by telling me that when they wanted to swarm they build queen cells, when they want to supersede they build supersedure cells and when they lose a queen they will build emergency queen cells.

Let me explain further. Firstly there are queen cells which are triggered by the desire to swarm. In its simplest form this is when the hive splits and divides. Any fertile egg laid by a queen will either produce a queen or a worker. In both theory and practice each fertile laid egg, given the right food, is a potential queen. The process of production means that each laid egg is initially provided with the same food. After three days the diet is changed for the workers; those that are to be queens retain the same diet. (I would like to think it is a chocolate and chip butty diet. Maybe that's why I look like a hippopotamus and lost sight of my waist twenty years ago.) As the bees fed royal jelly change into queens, the babies on their special diets become workers. If a colony suddenly loses its queen, the workers will select those larvae in worker cells that haven't had their diets changed. These will be fed the queen's special diet and this will result in a queen replacement.

So what does a queen cell look like? Well it's easier to look at the range of cells and try to clarify which does what. Let's look at play cells first. These are, potentially, queen cells, but play by name, play by nature. An old bee keeper with 70 years experience described them to me 'as being something for the bees to practice on'. I like the idea of the bees, all with their little reflective jackets and hard hats getting ready on their building site to construct a dummy cell before the making the real ones. The play cells look like a small acorn, with a downwards facing opening. What is noticeable is that the edge of the cup is turned inward and upward, like it has been peeled back, and it is thicker at its rim and

usually quite small in size. Think of these as a 'dummy run'. Keep an eye on these cells, but unless they are occupied just ignore them - there is no need to remove them, they aren't a problem. In fact I usually leave them on to show students what they are looking for. If you do remove them the bees will only start building more. The queen may choose to lay in the play cell, however. If she does, the attending bees will start to flare the sides of the cell outward and accentuate it, turning it into a proper queen cell. So if you see the cell being treated like this and possible new wax being placed on the cell, then it's worth a look.

Now we will look at swarm cells. These are in effect a large number of queen cells. It would be wrong to say that your queen will lay a specific number – it will undoubtedly vary. But what there will be for certain are queen cells laid over a period and therefore of differing ages. Even though the queen has laid a number of daughters it doesn't mean the hive is going to swarm. I always think when swarm cells are produced the decent thing a colony can do is at least swarm! The wishful thinking ends here. You will often have identified a hive that shows all the signs of swarming and you spend a lot of extra time waiting to catch the swarm or prevent it from running amok. Nothing happens, but when you turn your back, they swarm. A bit like when you are waiting in for that special delivery from the Royal Mail; you slip out and in a flash they call and "you were out". Tragic. Even when you think there is a swarm due, what can happen is that the other worker bees gain a scent of the newly emerging queens, rip open the cells and kill them. When you are looking for a queen cell, take nothing for granted. Obviously you will always look in the obvious places, the favourites being either on the sides of the

frames or on the base of the brood frames. But look in the less obvious places too: I've known them to be on the inner walls of the hive and under the roof. And don't be fooled by the size. Don't always think they are going to be large. Sometimes they can be quite small: they will follow the characteristics of a 'normal size' queen cell, but just look rather tiny.

Next, there are supersedure cells. This is where the honey bees replace the queen with an entirely new one without the usual death, destruction or replacement by division. The old queen may live side by side with one of her daughters without fear. She may continue to lay eggs. It is a moot point, but many consider that a hive will only supersede when the colony feels that the original queen is damaged or weak: this may be calculated by the strength of bees laid, the number of bees, or the inferiority or quality of those born. Think of supersedure cells as normal queen cells but with a different purpose. The cells have eggs in them that will turn into queens, but when or if this happens is down to the bees. One factor that might help you decide if they are supersedure cells is that these cells tend to be in the centre of the brood nest and not on the edges where you would normally find swarm cells. Only one of the supersedure virgin queens will survive; the remaining supersedure queens will be destroyed by the attendant bees. A point to consider is that a colony may build up and all the signs are that it is about to swarm, but then supersedes. It's one of those things that is best left to the bees to decide and for the bee keeper to watch and monitor. I know that sounds vague but I still find the swarming season one of those times that makes me feel anxious and I'm glad when July comes along and I can breathe a sigh of relief.

There are also emergency queen cells which are produced when the queen is lost. This can happen for a number of reasons. A common one is where the virgin queen flies to mate and doesn't return to the hive because a predator, such as a swallow, has eaten her. Over the years I have kept a record of the weather and alongside it a note of when certain migratory birds have returned. This record is not for predicting the weather like some old sage of the countryside, but to see when there will be a profusion of birds that might eat my queen when she flies to mate. If you have a long cold winter that stretches well into March or even April, then when your bees start to fly and the young queens are ready to erupt, you might have an unusual number of hungry swallows or house martins arriving simultaneously from South Africa to decimate your bees. If the weather is different over the winter period, the birds will arrive over a much longer period and this lessens the chances of losing a queen. Sometimes a queen is lost through an inspection. It doesn't matter how gentle you are, occasionally you can trap or even kill your queen as you replace a frame. Although I have personally damaged a queen on an inspection she has carried on happily, though slightly bent, laying eggs. Queens are tough but it's best not to be too vigorous. If a queen is killed then the remaining workers will burst into life and raise replacement queens as long as they have some young worker brood. Emergency cells can be laid shortly after the colony identifies a problem or the loss of a queen. Remember these are not what would be called primary queen cells - they were not designed specifically for queens to be laid in. These are used only when the queen has been lost for some reason (and it happens even with the greatest of care), so in effect they are converted worker cells made ready for a queen. Young larvae or eggs

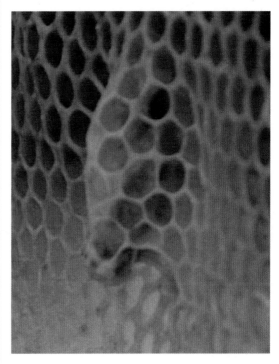

Queen cell spur

are usually in the cell already. The workers feed those that they have chosen to become queens rather than face the prospect of a queenless colony. There is a tricky part to this. If this scenario happens and you are trying to locate the 'new' queen cells, it can be difficult to distinguish emergency queen cells from 'ordinary' cells as they have not been 'purpose built', as it were. When trying to find these cells you may find they will be small and not necessarily located in the regular sites on the frames.

Other queen cells can also appear 'odd'. Sometimes you will identify a cell that appears to have been broken open. This might have happened for a number of reasons. I have often come across a hive that appears to have lots of lovely brood just waiting to burst out. There is a queen busying herself, laying eggs and yet there are open queen cells. Why? Quite often it is when a queen has emerged and the workers have killed her and sometimes they (the workers) tear down the queen cell to kill her. When the virgin eventually emerges, she will cut her way through the cell roof and emerge successfully.

If you are like me, when you do find a sealed queen cell you will probably feel a mixture of anticipation and excitement. (I always like to find the first cell and make a note in my records and the same with the last queen cell.) When you do find a sealed cell you should work on the basis that there is a baby queen in there. The process for the queen larvae is that when she is ready to erupt, the workers thin the top of the wax to allow easier access for the new queen. If you are able to see this process in action, it gives you a good idea as to when 'queenie' is about to make an appearance. Don't be fooled by looking at a queen cell and wondering why it hasn't hatched. If you have a queen cell that doesn't appear to be progressing, give it a very gentle squeeze at either side of the cell head. If there is movement on the cap it generally signifies that the queen has emerged; the cap head is hinged and the cap has fallen backwards, giving the impression of a very late queen cell (see image on page 64). Sometimes, however, queens do die in their cells (see disease section). You should be looking for a sixteen day period for the queen to be hatching, so record taking is the key here. I speak about this like a natural, but at times I'm the world's worst administrator; I would rather be out there with the bees - just ask Mole. If you are a little unsure and think that nature is playing tricks, or you think you might have got your dates wrong, make an opening in the side of the wax cell and take a peek - you can always reseal the cell wall thereafter if you are wrong on dates.

What should I do about damaged cells that are sealed? It really depends on the damage. For instance, if you find a cell bashed left and right but with the top intact, then it may have been an attack on the cell by other queens or workers. The queen may well emerge if the cell walls have been damaged but not broken, but you may well find you have a weak queen. If you find the top of the queen cell intact but there is a hole in the side, your queen might have made a side exit as the top of the cell was too difficult to work her way though and the walls were weaker. If you find torn or ragged cells, they have more than likely been ripped by the resident queen threatened by the presence of the other emerging queen. On occasions worker bees will reseal a damaged cell and then tear it down later. I don't know why. But don't forget, it may not always be a rival queen that damages your young queens. I know over the years an impatient or over-stung bee keeper can damage his queen stock by not being careful enough. I hold my hands up here.

Tips!

- **Record queen cell progress. Hatching should take around 16 days.**

- **There are 3 different types of queen cell, so be careful which you 'squish'!**

CHAPTER FIVE
ANATOMY

It's important that you as a bee keeper are aware of what a bee is made up of and what those parts do. I am sure that if I did a quick quiz and randomly asked the question "describe a bee's body part", no-one would say heart or antenna. They would automatically say the magic word - STING.

Let's look at the external parts of the bee first. You might wonder why I am doing this; it isn't a biology lesson. It is important, however, to recognise certain parts of the bee as abnormal or unusual traits may well indicate disease. The bee's anatomy is divided into three parts; the head, thorax and abdomen, or as bee keepers like to refer to it - top, middle and bottom. On the head of the bee you will see its antennae, the compound eye, the mandible and its simple eyes. The head houses the brain. The bee's brain, as you might have guessed, is a complex feature of this insect. The brain houses around 950,000 neurons which are individually specialised and manage to communicate successfully with the other surrounding neurons. The division of complex tasks created by the bee is purely down to the sophisticated nature of its brain and its functions. Working on this basis of specialized neurons, the complex functions that the bee performs would ordinarily require a much larger brain. It is a system of specialized nerves that allow the brain to communicate so successfully to other parts of the body. The bee has two antennae and has five eyes - three simple ones and two compound. Each compound eye has 150 primary cells allowing the bee to see polarized light and patterns. On the bee's head we would expect to see a mouth that the bees use to eat and drink. Each bee has a pair of jaws, a tongue, a labrum and a pair of maxillae. (This is about as technical as I shall get as there are numerous excellent texts you can refer to if you would like to know more.) The jaws will be used not for attacking prey as in many other insects, but for opening reluctant flowers. The tongue, or proboscis, is quite long and is used to seek

Smoking the hive can help prevent stinging

out sources of nectar from the most difficult of places. The labrum and maxillae are just like lips; their role is to support the proboscis when the bee is collecting nectar.

Now let's look at the middle or thorax. You will find the wings housed here. Every bee has two pairs of wings which are gossamer thin. The front wings are larger than the return or back wings, and a row of very fine hooks connect the front and back wings together so that when the bee is in flight the wings beat together. The wings are actually part of the bee's exoskeleton.

Let's now look at the bottom end or the abdomen. Needless to say the abdomen contains the heart, breathing tubes, stomach and other internal organs. I'm not going to go into the 'tib fib' sections of a bee's legs; suffice it to know they are hairy like mine and are used for collecting pollen, which mine are not. Their feet have a rather sophisticated pad and claw for manipulating

External anatomy of a bee

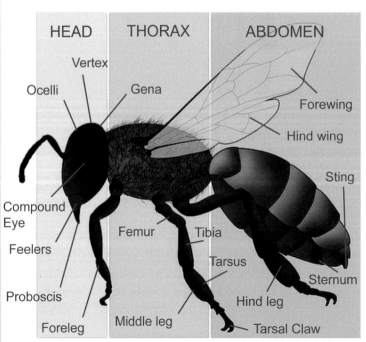

items. They also have a press for packing the pollen into sacks. Apart from the sting the bottom end has no real protuberances, so let's concentrate on the sting for a moment. This jolly item is a modified ovipostor or egg layer that has changed over millennia into a very effective weapon, although it does have a down side. When the sting is used the bee is generally dead within three minutes whilst the sting is still pumping away for a good twenty. Found within the sting is a poison sack and on the sting is a head like an arrow with tiny microscopic barbs that run at different angles. When the bee arches its back to attack, the sting is forced out along with the venom. The bee keeper has a cunning tool to keep the bees quiet and occupied so they forget about stinging you; smoke. When you smoke the bees they think that the 'forest' they are living in is on fire. They ignore the clumsy bee keeper and go off in search of food, mainly honey, to take with them in case they have to flee the hive and set up elsewhere. Because the only way a bee can transport honey is in their stomachs and then regurgitate it later, they have to eat as much as possible. When they have a full stomach they are less able to arch their backs and subsequently they will struggle to sting you. That's the theory anyway. It is believed that in prehistoric times, when bees were wasps, they would lay their eggs in meat and allow them to hatch. As bees developed and ceased to eat meat, sticking to a vegetarian diet, they retained their sting - I sometimes wish they hadn't. Some bees have developed to the point were they don't have stings, so maybe its just a matter of time and evolution and we can look forward to a future of stingless bees. Bumble bees, queen bees and wasps all have a smooth sting and can repeatedly sting to their heart's delight without any real effect – except on the recipient.

As for a bee's internal anatomy, well here we have it below, which is fairly self-explanatory!

Internal anatomy of a bee

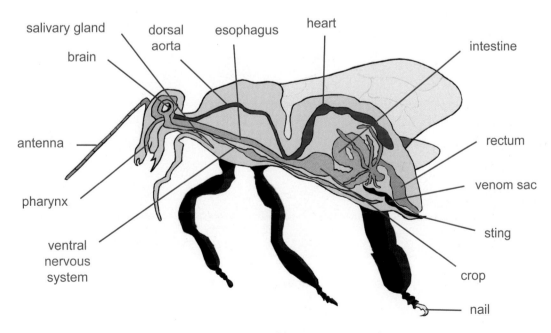

Managing your hive. The W.B.C. hive is pictured above.

CHAPTER SIX

HiVe MANAGEMENT

I don't want to drone (get it?) on about disease etc., but in this section I really want to discuss not just the management of the hives, because in many ways, they manage themselves. A tidy hive is a tidy mind, or is it the other way round? Although I have to say that by keeping your hives and the surrounding area tidy you will go a long way to suppressing disease and it will allow you to enjoy working with your bees even more. I'm a fine one to talk, but trust me, if you can lay your hands on an item and have plenty of working space then it just makes it all that little bit more comfortable and easy to return to.

This section will look at not just the management of a colony as this has mainly been covered elsewhere, but the additional peripheral aspects of management - where do I put a hive, what are the best practices and the dos and don'ts. Bee keeping is by no means an exact science but it has always been considered to be the science of farming. What a colony requires from the bee keeper is a sound knowledge of its behaviour. A good knowledge of botany, an interest in the weather, an understanding of how the hive reproduces and a good knowledge of your locality will all help too. This knowledge can be achieved in a number of ways; you can read a book, join a club and share a hive, or opt for the deep end and get your own bees. You will find your way and create your own direction. Find a system and stick with it, mainly because not all the parts are interchangeable, but also

because other factors may need to be taken into account. If you have a weak back, for example, then choose a lighter hive - never make your hobby a nuisance. Look at both your environment and personal situation and decide yourself. I must stress, though, a move in the wrong direction can lead to all sorts of problems. You will get lots of advice from others about the best process to follow. I will give you an example. I used to belong to a bee keeping club. I learnt a lot, but when I felt that it was time to make my own mark and push ahead I was treated like the anti-Christ. The reason? Because I wanted to use an American designed hive, the Langstroth, and because it was produced in polystyrene. Because it was not made of wood and was not a traditional hive

National hive

such as a W.B.C. or a National, it was not considered to be acceptable. I was looking at this from a commercial angle, so using either the W.B.C. or the National (especially the W.B.C.) would make it far too heavy to be moving around on the moors. I have no problem with the 'plastic' hives, as they are called disparagingly, and I love to see an old fashioned WBC in an English country garden. In fact I have one myself on the farm and one at the honey farm; in many ways it's what people expect to see.

So where should you site your apiary? You can keep hives almost anywhere. Do consider if you have a dog or cat. It is really interesting when an animal comes into contact with a bees' nest/swarm/hive for the first time. They race up to it and almost tear up the turf to stop when they realise what it is - isn't nature jolly? If the hives are to be placed in your garden, remember the positions of the sun and prevailing wind. Some sites will always be better than others, however. A site that is flat, well drained and is protected will always be better than an undulating one with boggy soil and which is open to the

Tips!

- *A good knowledge of botany, an interest in the weather, an understanding of how the hive reproduces and a good knowledge of your locality will all help your colony.*

- *A site that is flat, well drained, and is protected will always be better than an undulating one with boggy soil and which is open to the elements*

- *Raise your hive off the ground if the ground is grass/boggy or damp.*

elements. If the hive is to go onto grass that may be prone to damp, make sure it's lifted off the ground. A W.B.C. hive will come with legs, although some other types don't. You can either make a couple of pairs and attach them, or alternatively, place a pallet or a couple of old tyres underneath the hive. I know it doesn't look as nice or pretty but it is cheap and effective. If you want things to look a little more neat and tidy and less like the interior of a crow's nest you should consider a number of ways to present your hives. You can use two rails of wood about 4 inches off the ground and pegged on either side to stop them falling over or two collapsible tables with the tops removed to allow for circulation. One method I have used is to take the trunk of an old telegraph post and cut it so one end is flat and the

Pallets can prove useful in elevating your hive off damp or boggy ground.

other pointed. Ram the pointed end into the ground so that it is fixed and solid. Attach a wooden square slightly larger than the size of the hive to the flat end of the post. You can then easily balance your hive on top of this contraption and, as there is an overhang, it will prevent any unwanted rodents from paying you an unseasonal visit. Space your hive stands about six feet apart and this will allow you, if you are a clumsy oaf like myself, enough room to move around without causing too much damage or destruction. If you are choosing a system of tyres/pallet/frame stands then you must ensure that weeds and grasses around the hives are kept under control. Do not go around the hives with a lawn mower for obvious reasons and DO NOT use a weed killer to keep the weeds down in front of the hive. Even if the weed killer is described as bee friendly forget it as sometimes, although it won't kill the bees, it will upset them if the chemicals are too strong. The only sure way of keeping the weeds down to size is to cut them by hand.

Make sure that access to the hives is clear and free from obstruction so you can move your hive easily. You can use a hive barrow if your back isn't what it used to be. These things grab the hive at its sides and make it much easier to move. If you are using one of the wooden varieties of hives a hive barrow might be a good investment, but not if you have hives on a roof or bay window though!

Depending on the size of your garden you may decide to have more than one hive, but don't think your bees will stay in your garden alone; they will certainly search elsewhere for food, so there is a possibility of keeping perhaps three of four hives in your garden, on a patio, on a roof or under a tree. (If you choose the latter make sure it isn't too shaded or that branches are bashing against

the hive and annoying the bees.) Ensure that you don't draw any attention to your hive as some people will, for some unknown reason, want to vandalise your pride and joy. (This is one advantage of a polystyrene Langstroth hive – it doesn't look like a bee hive!) What about the availability of pollen and nectar in your area? Well, in the towns and cities you are generally better off than in the countryside. You are enabling your bees to have a varied diet which is far healthier than a mono diet of just rape seed or beans. It will usually take a year for you to establish whether your site is a good one. Obviously the weather and environment will play a big part, but by keeping an eye on your bees and seeing what the flow of honey is like, you should be able to identify the suitability of the site you have chosen. Remember when you do keep bees in your garden/on your rooftop etc. you will be responsible for every sting in the neighbourhood. Even if your bees are not the culprits your neighbours will believe they are. It is worthwhile to try and identify whether there are any other bees close by, whether they be hived or feral. The feral ones will be more difficult to identify, but someone might have noticed them. I said earlier to check the deeds of your house as you may not be able to keep bees and also check with your immediate neighbours as to whether they object to you keeping them. There is a very old and very good English saying: "good fences make good neighbours". Check with your neighbour to see if they are happy with you having a hive in your garden; the same goes for on roof tops and bay windows. Reward them with a jar of honey at the end of the year as they will feel they have been involved with the process as much as you. If you have had disputes with your neighbours about hedges, boundaries or fences, keeping bees may not be the easiest of ways of getting on with them. It's also handy to know what's in the locality in case of disease, either for you to be told or to tell your fellow apiarists. Good manners cost nowt (for our American cousins, nowt is a Lancashire word that means "nothing") and you may find that you can help each other out. I've never met a bee keeper that wasn't prepared to help out or offer the use of a piece of kit.

You do need to consider accessibility when placing your hive. How easy will it be for you to get to it? Something on a roof top or bay window might be difficult and may upset your neighbours if you have to pass by their window with your ladders every time you want to inspect your hive. So a hive in the garden or in a park sheltered from prying eyes may be better. Also consider the weather: roof top hives really need to be secured, as do ones on bay windows.

The chart opposite gives you some examples of how to place your hive, taking into account a little more detail by giving you three examples.

You could put forward a very good argument for all of them, with just with a little bit of tweeking.

Positioning really is the key to successful bee keeping. Let me give you an example. I once asked a hill farmer if I could place some hives on his land. He explained that a previous bee keeper had used this land and that he had no problems. It was a nice warm day at the end of the summer and in turn this attracted a number of fell walkers (trampers I think you call them in America). It took Mole and I from dawn until dusk to get the bees on to the heather, settled, unbunged and working away. After the weekend I received a call from the farmer wanting the bees

moved from the hillside and re-sited several hundred yards away. It may sound easy enough just moving the bees a bit further up the hillside, but moving them such a short distance serves to confuse them and caused terrific trouble for Mole and me. A day's job turned into a two day job: back at night to bung up the hives, back first thing next morning to move both hives and pallets and so it goes on. Apparently some fell walkers had wandered off the public footpath and on to the private land to have a look at the hives and were subsequently stung. It was their own fault but that experience cost me three days labour. So the lesson here is always to be exact about where the hives are to go, and for how long.

There are bee keepers in both London and New York who move their hives from one part of the city to another. This is because the councils all have a different programme for planting and therefore a different pollination period. These urban beekeepers are harnessing the properties available to them and rightly so. You may wish to approach a local park and ask about placing your hives there. I have hives in Manchester in an area where the hives are out of sight of the general public: out of sight and out of mind it seems as I have never had any trouble.

A hive in your own garden is certainly the ideal site, but please don't think I am only trying to steer you in this direction for you to tell me you don't have a garden. I can guarantee that if you ask around for the opportunity to place a hive locally in someone's garden, you will be swamped with offers. I know this for a fact because when I give talks about the joys of bee keeping, many, many people in the audience are keen to offer their land, however large or small. There appears to be a big

HIVE LOCATION	PROS	CONS	LOCATION RATING
The back of an east facing garden with a wall behind and a hedge to the right.	The hive would only need turning slightly to the south and towards the hedge. This would give the bees lots of sunshine throughout the day and keep them active and the hedge to the right would give the bees the obstacle they require to make flight quickly. From ground level the bees will need to get up to a good height quickly so as to avoid flying into your neighbour's property, or worse, hair. When they are up a good 10 to 12 feet they will be scattered and will form their own flight line back to the hive.	You're hive could be close to a neighbour's property, an area where children are playing or someone's washing, but you can guide the bees away from certain points by placing obstacles, like wicker screens in their way.	
An open roof with a school close by.	Well the school is neither here nor there. Your bees are so high up they will form their own flight line above the height of the playground.	Public/parental misconceptions. (See Myths & Legends chapter)	
The roof of a bay window.	The roof of a bay window is really ideal. It is high enough to get the height they need and low enough for you to do spot checks when you need to.	My only concern is that most bay windows are on the front of the house and having a hive in some areas of the city and towns can attract trouble from vandals or thieves.	

movement in society these days, where people want to get involved with the countryside, but don't really want to get their hands mucky. So encourage your neighbourhood to plant up bee friendly flowers that will keep the environment pleasant, will give you a healthy hive and will offer you the opportunity to produce a very local honey. In spring the urban environment will be 2 to 3 degrees warmer than in the countryside and there will be broken, if not diluted, wind patterns. There will be lots of bulbs for the bees to work on in spring, an abundance of flowers in parks and gardens in the summer and the likes of ivy in the autumn.

You could try some hives in your local park

How do the bees fare in an urban environment? Are bees more or less stressed by different environments? Certainly bees left in one place are happier. How do you measure a bee's happiness? Well you can't, but you can assess their stress levels via the varying diseases they suffer from. Bees that are not moved are likely to be disease free and because of that the bees are more content with their particular environment. If you are thinking of an out of town apiary site, bear in mind the pros and cons of having to travel. I once had a small number of bees on courgettes (zuccini) 30 miles away. At my farm it was belting it down, real cats and dogs rain, so I stayed in as there was no point in inspecting some hives, getting wet and spoiling the day. Five miles away it was cracking the flags, a beautiful day. So having hives too far away can certainly work to your disadvantage. I also feel that urban bee keepers are a class apart from commercial ones and the farmer who keeps one or two. I actually feel you are taking a radical approach and have that

warrior instinct – you are doing something that traditionally is not expected in towns. Well done. If you do decide to split things and try one a little further away from home - your park perhaps - you will be expected to manage it, take responsibility for it and fend off any one who gets stung. The hives I have in a Manchester park have been there for some time and the bees are quite happy. The groundsmen are aware of the bees and give them a wide berth, but at the same time have expressed an interest in them. They like to feel that they are working alongside the bees and have all noticed an increase in the fruits on the trees and shrubs. It's incredibly rewarding when a stranger thanks you and expresses an interest in apiculture. Aaahh, the tea has arrived – Assam this time.

If you are to move a little away from your usual stomping ground there is another factor to consider apart from flight paths and neighbours: you should not have too many hives for the location. I know bees will fly three miles but you will make it less stressful for the bees if you give them the opportunities to forage as close to the hive as possible. A hive in a public park will offer the opportunity for a variety of pollens and nectars and should produce a good honey, but remember an active hive will require at least ¾ of an acre to keep it going over the season. Even in a thirty acre park in a big city you won't find thirty acres of foraging material. In thirty acres I would put five hives over the year, monitor them and keep record cards of their success or failure. If you know there are 240 acres of rape seed or beans, then yes place 275 hives down, but that will only give you a build up and potential honey for the pollination period of around 4-6 weeks. After that you will have to find a home for 275 hives!

One question I can hear you asking is when and how will I get some bees. As bees are at a premium these days trying to source them can be difficult. You can approach a bee association - there are plenty of them and they can be found on the internet. It's always better to get them from as close as possible to where you live, however. If there is a bee breeder in your region you can order/buy a queen from them. I foolishly imported some from another county many years ago. It wasn't the fact they came from Shropshire, it was the fact that I had not done my homework because in Shropshire at that time there was a lot of disease around and I was bringing potentially diseased bees into a disease-free county. The bees were not diseased but I didn't know this at the time: I wasn't exactly Mr Popular.

You can buy bees on the internet these days but I would think carefully about this unless you are going to look at them. You can buy a queen and a small nuc and build up on that. You can also place an order with an established bee breeder and wait for your bees to be ready. One way of gaining

a colony is to place some swarm bait in an empty hive and hope that a passing swarm is attracted. This can work but is a little too unscientific for my liking.

When you do buy bees don't get too carried away with the thought of buying the first ones you see. Check their suitability for your location. If the bees have been raised inland and you live in a coastal area, will the increased salt levels upset them? Have the bees been raised in a system that you are already using, for example, are they on National frames when you are using Langstroth? This is not a major problem, but it all adds to the issues when trying to start your own colony off in the best possible manner. One question I always ask is about the temperament of the bees and what strain they are from. If they are from certain strains – ones that I know are aggressive or won't be happy in my area and climate - then I steer clear. Equally, have a look for disease, especially on the comb. Try to identify where these bees came from; are they from the seller's apiary or are they a swarm? Check for things like the strength and age of the top bars of the frame: you don't want the bar to give way and your precious cargo come tumbling down. The ideal situation for me is a nice small quiet nuc from a local bee keeper who can identify where the bees came from and moreover what their temperament is like. When you go to look at bees take someone with you who knows what to look for. And as for price, that's a hard one, because at the moment bees are attracting a high price and unfortunately some bee keepers are having their hives stolen.

Also consider how you are going to transport these bees. Car? Van? Or by courier? If you fetch them yourself you will need equipment for moving them, a bee suit in case they escape in your car, a smoker and gloves. And what if the car breaks down? The rescue services won't help if there are bees around.

Keeping records has been touched upon earlier but it is important, no matter how big or small your apiary, to keep an accurate account. You can design your own record cards, or you can buy them from a bee keeping supplier. A record card will help you to identify the nature of the queen, whether she has been crossed and with what and what the outcome was. You can also look at the numbered hive and then balance that out with the amount of honey you have drawn off. One thing my Uncle Nipper liked to do was to note the nature of the bees; were they angry, nasty, nice or docile and then look at whether there was a correlation between the nature of the bees and what crops they had been on. I find bees on rape seed tend to be aggressive. In addition, I always make notes on the weather, including the temperature and, importantly, the humidity. I would say that these days, with the desire for transparency, disease detection and an ever-decreasing bee population, it is important to help the bee inspectors as well as yourself. It's not enough just to point at a hive and give its history. I have found recently that the practice of keeping the record cards next to the hive in a waterproof poly pocket is ideal. Make sure that it isn't going to get wet and the bees can't gain access to it, though, as they will eat the card for fun.

If you are keen to start in bee keeping you will note that in the catalogues from bee keeping suppliers the equipment is expensive. Do not be put off by this; as with all hobbies your initial start-up costs are always high, but then your equipment will last for many, many years if treated correctly. Your first jar of honey will be the most expensive! I've kept my first jar from every season.

Now we come to the most important acquisition of all. Bees. You will no doubt be all excited and wanting to get on with getting hold of some as quickly as possible. Well let's look at the facts. In September you won't get any bees unless someone sells them to you. I would be a little unsure if someone offered me them this late in the year. You need to identify which bees are most suitable for your area. In addition you really could do with a whole season to get to know your bees; you need to understand their characteristics, their nature and productivity potential. It is much better to wait until the spring to get your bees.

When it comes to choosing your bees it's a bit like looking at your wife's mother to see what your wife will look like when she is older. Have a look at the bees you are about to buy and look at where they have come from. Certain bees have certain characteristics. I use a dark European bee, mainly a Scottish black. They are suitable for an area from mid-Wales right up the west coast of the United Kingdom. These bees have their own characteristics. I like them, but these particular bees have a tendency to be aggressive, or let's say, feisty. They also have the ability to over-winter well for a number of reasons. They have long hairs on their legs which keep them warm and they are good pollen gatherers. At the same time, however, they are slow to build up in the spring, although they do well in cold summers and are able to forage at quite low temperatures where a number of other breeds will struggle. I have seen these bees happily working away on the moors on the heather in rather chilly weather when other bees, mainly the Italian ones, are wrapped up in bed having a chip butty and a cup of tea. The Scottish blacks are not really known for their good housekeeping, however, and this can give rise to wax moth infestation and a tendency to allow varroa to infest. Don't get me wrong, they are not messy or slovenly bees, but others have a better approach towards housekeeping - I would rather have a sloppy bee that survives the winter rather than a clean set of bees that are dead in the bottom of the hive. These bees are also prone to nervousness which means that as you open the hive there is a tendency for the bees to fly and sting. They will come out to greet you even if you don't want them to.

Other bees are better suited to other climates. Italian bees or ligustica bees have terrific characteristic and are very popular around Europe, but have a look to see whether they will be suitable for your climate. In the UK it's not a matter of a north/south division, however, more an east/west one. In the South East and South of the U.K where the climate is generally warmer and dryer you will find that the ligustica bee is better suited. They are of Italian/Greek stock and have a fairly good, solid nature. They are a small bee, very gentle and a good breeder. They breed at an alarming rate so will need plenty of room. Remember, though, that although they might be good breeders they are not good foragers unless the weather is right. So lots of food/stores need to be available, especially in winter. They are easy to work, clean, very good at housekeeping but are generally prone to drifting and robbing. They are slightly multicoloured - not exactly a bright yellow, but "slightly tanned", although colour means little. I think prospective bee keepers have a tendency to associate the black bee with a more sinister disposition and the nice summery yellow ones with more pleasant characteristics.

The second most popular bee in the world

is the Carniolan bee (or Carnica as it is sometimes referred to). These are really nice quiet bees, especially the Slovak and the German breeds - the type of bee you could invite to dinner and your mother would be proud of your choice. They are steady and not prone to drifting, robbing or swarming and there is very little problem with chalk brood. They will over-winter well on low food/stores and on a small colony basis. These bees are a good all rounder and will breed bees of a similar nature. In the few colonies that I have had of these bees I noted (on my record cards, of course) that they were the quickest to build up. As soon as spring came they were away and breeding very fast. I've always felt these were the best bees to buy, although they can be expensive and sometimes near nigh impossible to find. They are a hardy bee and they are used to

Tips!

- **Scottish black are suitable for an area from mid-Wales right up the west coast of the United Kingdom. They are feisty in nature and weather well in the Winter.**

- **Italian bees (ligustica bees) have great characteristics and are very popular around Europe, but have a look to see whether they will be suitable for your climate.**

- **The Carniolan bee is a good all rounder. They are steady and not prone to drifting, robbing or swarming and there is very little problem with chalk brood.**

a cold climate and winter well. This is true, but the continental winter they are used to is a raw, dry cold, where temperatures may not see freezing point for months. The cold winters of the west coast of the United Kingdom are not like continental ones. One day it's snowing, the next it's sunny and the next it's rain and wind. A Carniolan may well struggle with this environment. Maybe the ideal would be to cross a Carniolan with a dark British bee? All countries tend to have a 'national' species of bee and a lot of bee keepers will support their own country's bee. I can understand that but when bees are dying out and the stocks of national bees are overstretched then perhaps its time to bring in some new blood? I have looked at Polish queens, Carniolans from Slovakia and bees in North Africa and New Zealand. French bees are nasty and aggressive - this is not meant as an attack on our Gallic neighbours - it is a wonder why they can not defend themselves against attacks by Asian hornets, because every time I have had dealings with French hives I have been mobbed and had a most unpleasant feeling from them.

I've mentioned various different types of bees in this book, some good, some bad, but you will no doubt have heard about the Africanised bee. The killer bee; the bee that would sweep the world and kill us all, if media reports are to be accepted. Yes, Africanised bees are aggressive and I wouldn't want to come face to face with them. But even in Brazil where they were introduced in the 1950s they have not had the same impact as their reputation would have us believe. (They have, since then, moved north and into the very southern parts of America.)

Their characteristics couldn't be farther from their European cousins. They are a smaller bee, and fast, tending not to divert from their chosen flight path and, most of all, an incredibly hard worker. You might think that that sounds like everything you want in a bee. Well it's the opposite in fact because they are incredibly sensitive to colony movement. They easily distinguish between wind in the tree and a hand trying to move them. Once they have been disturbed the alarm is raised and the majority of the colony erupts. The difference between the Africanised bee and its European friends is that the African colony will not only communicate quickly within its own colony but will communicate with other colonies close by. They will sting for the fun of it and if you have several colonies attacking you, even with the best bee suit in the world, I think you will struggle to survive. A mass attack on any one deemed to be a danger to the colony is the norm. Not all colonies are in trees or bushes; many are within gullies and large colonies can be found underground in sandy soil; step on it and you will know in a very short time. You can't out-run them and they will follow you, just like in a comic book. Your only escape may be to jump into a river or pond, like the comic book character – let's hope you can hold your breath for a fair length of time. Africanised bees can't be managed for commercial purposes and have a tendency to swarm like Nepalese mountain bees if they are disturbed too much. They won't just fly to the nearest tree and possibly come back when it's time for tea. They will disappear. I am unsure as to the effect Africanised bees might have on the native European bee (not that there is a danger of them crossing the Sahara), but there is a danger of them coming into contact with bees in America. American bees are descended from the European bee and I shudder to think of the outcome.

There are many varieties of bees across the world, all with their different characteristics and some will fit your needs and many others won't. But a bit like any one with a keen interest and taste for their chosen subject you will shortly be able to distinguish parts of a hive, types of bees and the flavours of honeys.

I wanted to mention the issue of drifting and then close this chapter on bee and hive management. Drifting occurs when one bee leaves a hive and ends up in another. This can happen when hives are in a line and there is a strong wind. Is this a problem? It can be but you will often find that if there is a sufficiently good flow of honey and an 'alien' bee arrives with a stack of nectar then the bees will allow her in without asking any questions. This can result in one hive dwindling whilst another is quite buoyant. (My first thought in this situation would always be towards disease, but then I'm a little paranoid about it.) 'Unbalanced' colonies are of no use to either amateur or commercial bee keepers, so you can try one or two ways of preventing this. Try not to place your bee hives in a line, but arrange them in a pattern. Painting the hives different colours can also be a good idea so that the bees 'home in' on their own hive. Avoid reds, but white, yellow and all shades of blue are fine.

CHAPTER SEVEN
EXTRACTING THE HONEY

Well here I am with these boxes of sticky mess that can be sold, eaten or swapped. (I like the medieval system of bartering best - a bee keeping friend of mine swapped honey for her Christmas dinner once: the turkey took quite a number of jars, but she got two dozen mince pies for half a pound of honey - a real bargain.). Whether you are a commercial keeper, a large scale amateur or have a single hive, you will still have to remove your honey from its comb. Although this can be done in a variety of ways the principle remains the same. From an amateur's point of view you will still need to comply with certain rules and regulations to ensure that your product is safe and suitable for human consumption. Take the scenario: you hand press the honey out of the comb and bottle it. Yummy. But I can guarantee the honey will be full of bee bits - heads, leg, wings and stings. Although this may appear very natural, all you need is someone with a sympathetic solicitor to sue you when they eat some honey, get stung on their tongues and panic.

Well where shall I extract the honey? Can I do it in my kitchen , garage or outside in the warmth? You can extract easily if you meet the criteria laid down by your local or regional food inspectors, although in the

UK you should comply with honey-based legislation, which can be found under the Honey Regulations for England and Wales, or, if you are in Scotland, there will be regional Regulations. All of the regulations are mainly European based and cover a wide area, with various aspects being vetoed and isolated for each nation state. In general, though, the wording doesn't alter that much. In America I would expect you to be covered by State Law. If you look at the food hygiene regulations for the UK all food preparation must be carried out in hygienic, sanitary conditions. You will need to contact the Public Health Department if you want to produce honey for human consumption and sale. If you are giving it away for no reward then you can present it in whatever fashion you wish as the person is accepting it as seen.

If you are preparing your honey for commercial purposes then your extraction environment must obviously be clean. I'm not going to go into what constitutes clean as I believe a lot of it boils down to common sense, but you should look at the regulations to see what is required. Obviously, where you extract your honey you must have proper washing facilities. At my extraction unit we have installed a shower - we don't have to have one of these by law, it's our choice. But you will have to have at least hand washing facilities and you must have a separate sink for washing items other than hands. As well as a hand wash basin (or access to one) you need clear lighting and access to a fresh air supply. Remember you are usually going to be extracting honey in a closed room which is warm and free from draughts, otherwise you will find that the viscosity of the honey will slow down and it will be harder to extract. Whether you are pressing or spinning, it will need to be warm. Although you can

warm the honey, at all costs avoid boiling it because you will destroy the natural benefits found within. You might find the odd person that wants their honey to be cold extracted because honey starts to lose its benefits when subjected to too much heat. If you are cold pressing honey then the filtration is much more difficult, especially for the likes of heather honey. Bear in mind the dangers of a cold pressed honey, however. Some honeys, especially the likes of rape seed, will set like rock very quickly. You are aiming to get the honey off the frames and have them back out wet for the bees as quickly as possible so you have to extract very quickly – with quick-setting honey you will undoubtedly have to keep the frames warm. If your extraction programme warrants it you can build a small area to keep your frames warm and ready for extraction. A small cupboard heated by a fan heater would be ideal. You can take your boxes of frames, place them inside and take the temperature up to the point where the honey starts to drip. If they appear to be heating up too much then rather than turning the heating up and down, try spraying the combs with water. This will allow you to maintain just the right temperature.

Ideally, your extraction area will also be constructed of material that is easily wiped. Now this doesn't have to be expensive plastic boarding, but it does have to be permanent and non-porous. So a wall or surface that gets splattered by honey needs to be able to be washed down with warm, if not hot, soapy water. If you can guarantee that a plastered and painted wall would not flake or chip it would be suitable, but I have yet to find one that doesn't! What is suitable and relatively cheap is the glass fibre sheeting you can get from agricultural suppliers that is used in milking parlours. This sheeting is dipped in water and then hung like wall paper using gloves to hang it and smooth it out. It can be moulded into the corners and the creases in the floor and walls. It can be wiped and washed down and comes in a variety of colours, but I would stick with a nice white. Whatever covering you use, though, needs to be smooth and non-porous so that there are no areas that will harbour grease or muck or allow other visiting insects to set up home. You will have lots of bees in and out of your honey house and it can be difficult to keep other insects out at times as they will be attracted to the honey. Your honey house will also need to be served with the likes of wipeable table surfaces and stainless steel instruments. Extraction tanks and bottling equipment should all be stainless steel and you should have at least one table with a metal surface that you are able to wipe.

So much for the extraction environment. Now what about the extractor - you! As for your own appearance I would say most of that boils down to common sense. Clothing that can be easily cleaned, light in colour and that won't attract dirt or harbour dust or spoil is down to choice, but I've always found a pair of tough, white overalls do the trick. Sometimes when it is really busy and hot

I wear shorts and a tee shirt. And sensible boots. (Certainly not sports shoes or sandals – you're not on the beach!) And preferably boots that are over your ankle, wipeable and, most of all, with a steel toe cap. For extracting honey, you ask? Believe me, if you have ever dropped a bucket of solid rape seed honey then you will be grateful for the toe caps! You will also need to cover your head – you really don't want to find hair in your honey.

You must ensure that your premises are rodent- and pest-free. No matter how meticulous and clean you are, wherever there is suitable food - in this case honey - undesirables will be attracted to it. Deal with them and dispose of them sensibly. If you are using your kitchen, and as an amateur this is more often than not the case, be aware that you may end up sharing the space with some unwanted guests. When I was an amateur bee keeper I extracted in the kitchen and unfortunately left some honey uncovered. This would not normally be much of a problem except that my wife and I went away for a few days and when we came back the kitchen was heaving with ants. She nearly filed for divorce!

If you are using your garage to store supers with frames before extracting you will undoubtedly attract not only your own bees but you may find the local wild bee population and, even worse, the local wasps (yellow jackets) attracted to them too. The latter can be a real pest, not just because they sting for fun but because if you have hives close by, possibly in your garden, then they will search out this additional food source and start to raid and rob. And that becomes another issue entirely. I have a separate section to my unit where I can leave the supers away from the extraction unit, but covered and away from the sun. When the supers are ready to be placed in the machinery they are then taken into a separate room that has wipeable surfaces and stainless steel machinery. The important factor is that when the supers are off for processing they should be kept in a room that is secure. I don't mean that there should be security guards on the doors! Secure in the sense that the doors fit and there are no gaps and that any windows are screened to prevent unwanted guests. You will be surprised what an effort even the smallest fly will make to get to your honey: remember honey = hard work and you don't want to be giving it all away. I have doorways that don't have an ordinary door as it is difficult to open them and get a trolley through at the same time, so our openings have 'plastic strip' doors. They keep most insects out and you can either wipe them or power-wash them to get any soiling off. I find these are much better than a well-closing door. If you are going to put a door in your special extraction space, remember to make the opening wide enough to get through whilst you are holding a super, otherwise you can guarantee you will scrape your knuckles and drop your super.

You should always have very good, strong lighting where you extract. It makes sense not to be struggling in dim lighting when you are dealing with hot wax and hot honey, hot equipment and, in some cases, sharp equipment. A cut finger, no matter how small the bleed, will contaminate your honey and there is no way you can subsequently sell it. So safety is a priority.

So how do I start the process of getting the honey from the frame to the jar and then on to my toast. You should start by de-capping the frames. They will have a line of wax caps on top of each cell. If they have not then the honey is considered to be unripened and may be prone to fermentation. This is fine if you want to make mead but not for honey purposes. You will need a knife with a sharp, flat, smooth blade. This is better than a serrated blade as there is a habit of treating the movement like a saw and then you start to cut deeper into the comb. I have a heated blade which you plug into the mains and allow to heat up to a pre-set temperature so there is no need to mess around finding the perfect temperature. You can even get a steam knife which works on the same principle, but never having used one I can't comment on their efficacy. If you have a decapping tank, so much the better, but otherwise the position to take is to have the frame flat against your arm. Starting at the end nearest your body, place the blade almost flat to the wood/wax division and move away from your body, keeping the blade flat. Don't EVER pull the blade towards you – one slip and you can injure yourself very badly. You will see the honey being exposed and the caps rolling away from the comb. Now you are going to have to have somewhere for these caps to go: the floor isn't a good idea for obvious reasons, so have a large container to hand. You will be surprised at the number of wax caps that come away. Don't be tempted to throw them away because they are useful; they will still have some honey in them and we want to use them later. I use a tank that has a perch on top for the frames to rest on and can be operated by two people. It (the tank) is kept warm enough to allow the honey to separate from the wax caps but not too hot to allow the wax to melt. There is a valve at the base of the tank that has a filter on it and the honey passes through to the collection tank or bucket without the wax falling through. The wax cappings can then be used for foundation later.

Once the frames have been de capped they will need to be 'turned'. Can I not just crush the frames of honey and heat them up, you ask. Well you could, but I'm afraid that it's a fine line between melting the wax and separating the honey. If you have even a single hive then you are better to invest in a small hand operated extractor. These will take anything from two frames to several dozen, stand upright, have lids to stop splashing and a hand crank. They are either metal or plastic, but I find the plastic ones better because they are cheaper and easier to clean and maintain. The only downside to these extractors is to be found in the words "hand operated". It can be a long process – up to three hours - to extract all your honey so be prepared for magnificently well-developed biceps in one arm and an under-developed weedy second arm. I have seen a power drill turned upside down and attached and then set to run and that works a treat. You can get extractors that take both deep and shallow frames, which is perhaps preferable. When you come to purchasing your extractor, take advice from someone who has owned one – someone at your local bee club and the seller perhaps. My Uncle Nipper never owned an extractor himself but was a member of a sympathetic bee keeping club. They (the club) owned their own, as do many bee keeping clubs, and for a small fee he could borrow it. This made it a cheaper option and meant that the honey wasn't hanging around waiting to be extracted and attracting all sorts of unwanted guests.

Extractors fall into three types: Parallel Radial, Radial and Tangential. Radial extractors manage to spin the honey from both sides at the same time, whereas the tangential type only extracts from one side at a time in which case you will have to turn the frame around to do the other side. In the tangential extractor there is a mesh screen that supports the inner side of the frame but isn't terribly suitable for some frames. The tangential, although relatively good for the job, does have the impression of being the 'Betamax' of extractors – it does a good job and is a good design but is just losing out to 'VHS' versions because of one or two functions it's lacking.

One thing that I don't like about the tangential extractor is that because it operates on a single side extraction you find the process weakens the wall of the comb and it becomes heavy on the unextracted side. This can cause the comb to buckle and collapse so just when you had all that nice extracted honey collecting in the bottom of the tank you now have a lumpy mess in the bottom that will need extra filtration. But these are only my findings and I am a clumsy lump – you may have different views. You will no doubt find that the extractors at the clubs are suitable for your needs.

The next step on the road from comb to bottle is to strain. You really don't want lots of stings, wings and heads floating in your honey. When the honey comes from the tank or container it is still warm so it should now be passed through a metal strainer. I say metal because a cloth one will only clog up. I use a medium gauge strainer in the first instance to extract the larger bits and bobs. As the honey passes through you should then pass it through another, finer gauge strainer, either metal or plastic, so that you can be sure all the detritus is removed. You will also find that if you are working on a larger scale and you are running the honey into a settling tank that there will be a certain amount of scum. This will need to be removed before bottling and storage. Personally I like to bottle straight away, but I think the jury is

metal for the lids and you will need a label that conforms to the Trading Standards legislation. This label will have to have a sell by date and its weight in either metric alone or imperial and metric. If it is both then the metric must come first and the weights must be clear and evident to the purchaser. Your logo must not obscure any of the important pieces of information. You might wish to place a 'tamper-proof' strip on the lid and attach it to the side of the jar to show it has not been interfered with. This is not imperative in the UK – only desirable. If you decide to sell it wholesale then it's a question of what you will store it in, and the labelling. These days bee keepers generally use plastic containers. I do - they are really simple to use, easy to store when not being used and are far superior to metal ones as they never deteriorate. If you sell to a restaurant, shop etc., you must put on the weights and a description of what you are selling, but there is actually no need to say what your company's name is, although you must leave contact details. In essence honey doesn't go off so long as it's kept at a low temperature and preferably out of strong sunlight. The honey found next to Tutankhamen in the 1920's excavation in Egypt was in better condition than the boy king! After 4000 years he was in a sorry state, yet the honey was identifiable as such. Had it been left in a glazed, non-porous dish instead of an unfired terracotta bowl it might even have been safe to eat. If you are going to keep any excess honey in your house, a nice permanently cool environment is better. A garage is good except for when the temperature increases in the summer. The same is true of putting it in the loft. If

out on this one. It's a matter of choice. One advantage of bottling immediately is that the honey is of a good viscosity and will flow easily. If you bottle later you may have to heat the honey to get it to move for it to be bottled. I just feel that a second subjection to heat is unnecessary.

Storage space may also be a problem. If you are leaving your hive in one place you should, in an average season, make 40-60 pounds in weight of honey; if you are moving your hive around you could be fortunate and get 100 pounds. If you do get that much it's an awful lot of jars to store. Better, perhaps, to leave the honey in buckets or another suitable container to bottle later. If you are bottling for resale, remember that glass is an expensive commodity these days as is the you are struggling to sell all your honey, contact your local commercial bee keeper - they will usually be interested in taking it. I always do.

A word about heather honey. It doesn't react in the same way as other honeys. It is thixotropic and therefore will neither set nor run; it has an almost jelly-like consistency. If you want to extract heather honey it needs to be pressed; this means you will have to warm it slightly, lay the frame flat and exert pressure, a little like pressing grapes for wine or olives for oil but with less vigour. Heather honey will have to be at a temperature warm enough that it runs and can be filtered and collected, but not 'cooked'. You can loosen the honey in the frames by using an agitator. This is much like a rolling pin with sharp spikes on it which must be carefully washed after every use. You roll the agitator over the top of the frame, so that the caps are broken and in most cases the walls of the cells are too. Be very careful as these pieces of equipment are very sharp, heavy and have a tendency to trap old wax and honey. Once you have rolled the spikes over one side of the frame, do the other side. Then press or place in the extractor. If you agitate and extract, although you might not get all the honey out and you will have destroyed some of the cells, you can place the frame, wet, back into the hive. The bees will clean the frame for the honey that is left in it and will, over time, repair the cells. Nevertheless, I feel the best way to get the most heather honey from the frames is to press, but if you do press, the comb will certainly be destroyed, so both options have their benefits.

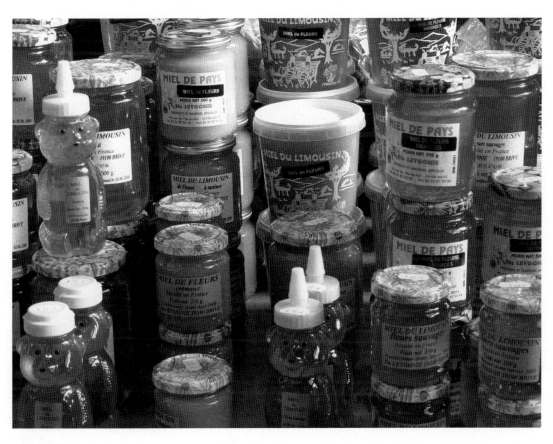

CHAPTER EIGHT

URBAN GARDENING FOR HONEY BEES

by Maureen Little

Bees need flowers to survive. Bees look to flowers to provide them with nectar and pollen - without these two vital sources bees cannot stay alive. But not all flowers need bees and not all flowers provide suitable food for bees. A beautiful, red, hybrid tea rose, for example, a quintessential flower of the British garden, is as useful to a honey bee as a pine forest is to a cow. A patch of flowering marjoram (Origanum vulgare), however, will have bees stacking like little aeroplanes just waiting to land and feed. As a bee keeper, therefore, it is essential to know what types of flowers are most suitable for bees - without an appropriate food source your bees will surely die.

This chapter will point out what plants you should be looking out for in the garden, parks and general locality, but it is not an exhaustive list - I have had to be selective and for the most part the ones I have chosen to include have been proven to be good 'bee plants', either in terms of pollen production or as a valuable source of nectar. If you are an experienced or fledging gardener yourself, there will also be advice on what you can do to help provide food for your bees. These topics will be looked at seasonally, but the scope of this chapter is such that it will give you an indication of what gardening jobs can be done. It is by no means a complete or comprehensive guide. There are numerous gardening books available, most of which will offer excellent, detailed information.

Gardening in an urban environment has much in common with gardening in the suburbs or in the country (just as keeping bees in a town has much in common with keeping bees elsewhere). This is certainly true for the 'why' of gardening. Most gardening, whether it is for growing vegetables, providing a pleasant place to sit, indulging in the growing of a favourite species of plant or offering a garden friendly to bees, has many shared tasks. There are two main differences, however, when it comes to the 'where' of urban gardening - climate and space. Firstly, climate . In a town or city, open spaces - no matter how large or small, whether public or private - are enclosed. This may seem contradictory, but invariably a garden, backyard or park will be surrounded by a boundary, usually in the form of buildings, walls, fences or hedges. This means that the enclosed space will not only be protected from the worst of the weather, but that the temperature is often 2-3°C (and can be as much as 6°C) higher than in the countryside . These factors provide a microclimate where the urban growing season can often be warmer and longer than in rural situations. This will have an impact on what plants will thrive and also on the length of the 'bee' season.

Secondly, space. There are two types of space within an urban environment: public and private. There is usually more public space in an average town or city than in a corresponding measured area of countryside.

It is true that large parts of the countryside are accessible to the public but they are usually privately owned. In towns, however, parks, pedestrian areas and even roundabouts are municipal, owned and maintained by the local council or other civic body. If these areas are landscaped then decisions on planting are made by the relevant amenity department. Often the planting is designed to be low maintenance (lots of shrubs) or very colourful (annual bedding) with, arguably, little thought to wildlife - and especially bee - support. As we shall see later, there are numerous shrubs and annuals that meet all the criteria. If, therefore, you find that your 'public' planting is not bee-friendly then petition your council to make it so. The vast majority of public areas in towns and cities are very well maintained but there are some neglected or abandoned spaces that have become the focus of 'guerilla gardeners'. These are ordinary members of the public who clear a piece of derelict ground and plant it up to provide an aesthetically pleasing and environmentally enhanced place. It is an illegal occupation and I would, accordingly, in no way condone it.

Private space in towns and cities is often limited, although a large number of houses do have substantial gardens, both front and back. Other residences may have smaller gardens, balconies, space for window boxes or can even accommodate a roof garden. Whether you have a large back garden or a simple window box you can still provide forage for bees. Before we look at some general information about bee-friendly plants I would like to make a general plea for gentle chaos in your garden (to paraphrase Mirabel Osler). I don't advocate letting your garden go, allowing nature to take over - even apparently 'wild' and 'natural' landscapes are carefully managed - but I would encourage you to be less fastidious about certain aspects of garden maintenance. The lawn, if you have one, for example. No matter how many times you weed and feed, scarify and roll, cut and rake, the average garden lawn will never reach the exacting standards too many of us try to make it achieve. Why not set the standard a little lower and the cutter blade a fraction higher to allow a few daisies, clovers and - dare I say it - dandelions to flower? Dandelions (Taraxacum officinale) are high on the list of bee-friendly flowers. Why not also tolerate a few wild flowers in the perennial border? Field scabious (Knautia arvensis) and meadow clary (Salvia pratensis) are every bit as good as some cultivated varieties and if you don't want them to seed themselves everywhere then by all means dead-head before the seed ripens. If you have a herb patch or grow herbs in pots or in a window box, allow some of them to flower. You may lose the delicate pungency of the leaves but it will be worth the sacrifice because an amazing number of herb flowers are superlative bee plants. I am thinking of marjoram (Origanum vulgare), hyssop (Hyssopus officinalis), mint (Mentha sp), sage (Salvia officinalis) and thyme (Thymus sp) in particular. And then there's traditional summer bedding: regimented straight rows with vivid, contrasting colours. You either love it or hate it. Bees don't mind straight rows but they do care about what flowers are in those serried ranks. We will be looking at what types of flowers bees prefer, and what colours they can detect shortly, but there is no reason why you can't have straight rows, vivid, contrasting colours and still be bee-friendly.

If you wish to grow some plants yourself, then as well as looking out for bee-friendly ones you must remember that not all plants require the same growing conditions. It is

worth looking carefully at the area you want to plant before you buy anything. Is it in sun or shade; is it exposed or sheltered; is the soil free-draining or consistently damp? All of these will affect what plants will grow successfully. Have a look at what plants thrive in your neighbours' gardens. If you are surrounded by camellias and rhododendrons then the soil will be acid and there will be no point in trying to grow pinks or other lime-loving plants. Look carefully at seed packets and plant labels or seek advice from your local nursery to help you decide which plants to buy . Planting the right plant in the wrong place could prove both disappointing and expensive!

If you don't have space to plant anything in the ground you can still grow things. If you have a flat roof you may think of creating a roof garden. More and more people are turning this 'dead' space to a truly living one, particularly in towns and cities. (You must, however, consult a structural engineer before you start bringing tons of soil up the back stairs, otherwise you may find yourself staring at the remains of your herbaceous border instead of the television one evening.) In the same vein as a roof garden is the 'vertical garden' or 'green wall', summed up in the term 'hortitecture' (a hybrid of horticulture and architecture). This is where a growing space is made on the side of a building or other vertical structure and planted up with suitable vegetation. Again, unless you have the technical expertise yourself, call in the experts for advice. If a roof or vertical garden are out of your scope

(they are certainly out of mine), you can still have some pots or containers in a small, paved backyard. A balcony too is an ideal space as long as it can support the weight of the pots, and even a modest window box or hanging basket can sustain a succession of bee-friendly plants.

There are three main disadvantages of growing in containers, however. The first is the lack of physical space which will preclude growing any very large or deep-rooted plants. The second is watering. Moisture will evaporate far quicker from a container than from open soil and during the summer months your pots will undoubtedly have to

Lavender

be watered every day and sometimes twice a day. As a rough guide, 2.5cm (1 inch) of water will penetrate about 15-20cm (6-8 inches) of soil, so you can see that even if it rains you will still have to top up your containers. Thirdly, the plants will soon use the available nutrients in the compost and because their roots are restricted they cannot search down into the soil for food. You will have to provide food for them in the shape of slow-release fertilizer pellets or by adding plant food when you water. In any case, follow the recommended application rate on the packet - too much food is almost worse than not enough. There is a major advantage in using containers, however, and that is you can choose your growing medium. This means that you can have neutral, acid or alkaline soil; free-draining, moisture retentive or boggy soil, which in turn means that you can grow whatever you choose and are not restricted by the type of soil you have in your garden.

Let's now look at some more bee-specific topics. Bees are active and will forage for pollen and nectar from early spring (March-ish) until the first frosts (often in October, but sometimes not until November, or even later in towns and cities). This essentially corresponds with the growing season of plants which isn't merely a coincidence - nature designed it that way! Bees - like humans - like a varied diet. Surprisingly this is less of a problem in an urban environment where there are no great swathes of oilseed rape or field beans that are found in an agricultural landscape, but instead there are lots of smaller private and public spaces which together make up a vast food source.

The nature of the plant itself is worth bearing in mind. The nearer the cultivated plant is to its wild, 'natural' original, the better it

Tips!

- **Enclosed garden spaces are likely to have higher temperatures than those in rural areas so select your plants accordingly.**

- **Try not to 'over garden' your garden!!**

- **Allow some of your herbs to flower.**

- **Water your pots/containers at least once per day.**

- **Bell shaped flowers also provide a good source of food.**

is for bees. Not all flowers are pollinated by bees (some are pollinated by other insects, some are wind pollinated and some reproduce vegetatively), but if we consider that the main purpose of just about every living thing on our planet is to reproduce, then anything whose reproductive organs have been interfered with will be unable to procreate. This is especially true of plants that rely on bees for pollination. Highly bred or hybridized plants often have double flowers, which render them sterile. The 'extra' petals of a double flower are in fact a genetic mutation of the sexual structures of the flower which means that there are no pollen-bearing male stamens and often no nectaries. If a bee or other insect does land on a double flower to look for nectar or pollen it will have had a wasted journey and will have unnecessarily used energy as well. Double flowers look good from a human perspective but they are of no value to bees.

The shape and form of flowers is also a good indicator as to whether bees will like them. Spiky stems with lots of smaller flowers like lavender (Lavandula sp) or purple loosestrife (Lythrum salicaria) are good bee plants; the bee can visit scores of individual flowers on one stem using very little energy and pollinating numerous flowers at the same time. Single 'flat' flowers such as those of the daisy family - Michaelmas daisies (Aster novi-belgii and A. novae-angliae) and black-eyed Susan (Rudbeckia hirta), for example - where the pollen and nectar is easy to access - are ideal, and there is a 'landing platform' to boot. Bell-shaped flowers too, such as foxgloves (Digitalis sp), will provide a useful source of food, although some are better suited to bumble bees rather than honey bees.

Bees see differently from us. Humans can see all the colours of the rainbow from red through to violet. A bee's vision is slightly different in that they see less than us at one end of the colour spectrum but more than us at the other. They cannot see red but they do see ultraviolet light, which we cannot. As you can imagine, this has a direct influence on which flowers bees are attracted to. They will automatically target flowers which stand out to them. The beautiful red rose that we mentioned earlier is all but invisible to a bee. That does not mean, however, that all red flowers are not appealing to bees. Think of the field poppy (Papaver rhoeas) - probably one of the most vivid reds in the flower world. Look carefully at the centre of the flower where the 'business area' is located and we see that there are dark blotches. This is the bit that the bee sees and in fact the pollen of the field poppy is dark blue.

Bees are particularly attracted to yellow and white which reflect ultraviolet light but don't restrict your planting scheme to just those colours - variety is certainly the spice of a bee's life!

Bees also like groups of the same plant. This doesn't mean that they won't visit single specimens, but they are more attracted to blocks of the same shape and colour. Blocks are easier to detect and when the bees do start to forage, they don't have to move very far from one pollen and nectar source to another. You shouldn't give your garden over to great swathes of the same plant, however - bees don't like a restricted diet any more than humans do - but perhaps you could group three of the same plants together instead of dotting them around the plot.

It's time now to examine each of the seasons and see what flowers are available, and what jobs can be done in the garden. The boundaries between each season are not definite: the margins are blurred and just because we theoretically enter autumn on the 1st September or some other arbitrary date, it doesn't mean that all the summer flowers will curl up and die. There is bound to be some overlap, especially if the weather decides to do strange things like temperatures of 20°C plus in April and then monsoon-like weather in July. So treat each section as a recommendation rather than a prescription. We could start with any season, but since Craig has started his bee keeping year in September we shall follow suit with our gardening year.

SEPTEMBER TO

NOVEMBER

I can't think of autumn without Keats's poem coming to mind. Autumn, that 'season of mists and mellow fruitfulness,' providing:

... more,

And still more, later flowers for the bees,
Until they think warm days will never cease;
For Summer has o'erbrimm'd their clammy cells.

The bees have already feasted on the summer bounty and now autumn presents the dessert: not as opulent as the main course of summer perhaps, but still worthwhile. As the days get shorter and the temperature falls the bees will take every opportunity to forage for pollen and nectar in order to fill the comb to the brim with honey reserves for the winter. (And then we come along and remove the honey - and very nice it is on buttered toast in the morning. But if we take the honey then we must replace it, otherwise the bees will starve over winter. Craig covers this elsewhere in this book, of course.) There are still flowering plants that we can look out for during the autumn and jobs that we can be getting on with in the garden.

Autumn Plants for Bees

Autumn flowering trees are, it has to be said, few and far between. There are no indigenous trees in flower at this time of year, but there are some non-native ones that are less widely grown and are very good 'bee' trees. These include Eucryphia glutinosa which carries large white flowers into late summer and early autumn, the pagoda tree (Sophora japonica), and the appropriately named Chinese bee tree (Tetradium danielli). The latter, especially, provides a profuse nectar supply when other sources are waning.

There is a handful of shrubs that are worth looking out for too. Fuchsia (Fuchsia magellanica), some varieties of hebe, Russian sage (Perovskia atriplicifolia), and viburnum (Viburnum tinus) (opposite - bottom) will provide some pollen and nectar. Ling heather (Calluna vulgaris) will also flower into the autumn. For more information about heather, see the 'summer' section). One of the most useful bee shrubs to have around at this time of year is undoubtedly ivy (Hedera helix). This flowers at the very end of the season and is an extremely important source of nectar, containing about 70% glucose. This is of particular value if we have good weather in October and November and the bees are still foraging.

Many summer-flowering perennials, if they are dead-headed, will put on another, albeit less flamboyant display during the autumn until the first frosts. In addition there are those which save their best until now, like Michaelmas daisies (Aster novi-belgii and A. novae-angliae), dahlias (but make sure they are

the single-flowered varieties), black-eyed Susan (Rudbeckia hirta) (opposite - top right), and valerian (Centranthus ruber). Another useful source of autumn food, and one which may not spring readily to mind, is the autumn-flowering crocus or meadow saffron (Colchicum autumnale) (opposite - top left). The leaves appear way back in spring and have died down long before the naked flowers reveal themselves.

Autumn Jobs for Bee Keeping Gardeners

The days are getting shorter, the bees are less active and the garden seems to be winding down after the hectic ostentation of the summer months. There are still jobs that the gardener can be doing in autumn, though.

One of the main tasks is to plant spring-flowering bulbs. A friend of mine who has a large garden (lucky person!) planted several thousand daffodil bulbs a few years ago. The first spring they flowered well, but thereafter the flowering rate diminished until this year all that appeared was leaves and a few scrawny flowers. She was bitterly disappointed. She had bought the bulbs from

a reputable source and had done everything according to the book, except for one thing. She hadn't planted them deeply enough. Instead of planting them two to three times their own depth she had just pushed them into her lovely loam soil and scraped a bit of earth over the top. It was a costly mistake in terms of time, effort and money. If you think you are planting too deeply, you are probably planting just deep enough.

As well as bulbs you can also plant out some spring and early summer bedding, such as sweetly scented wallflowers (Cheiranthus sp). Although you can raise these from seed sown in the spring, they are widely available at nurseries and garden centres. Don't bother buying them in pots. They are much cheaper bare-rooted and will come to no harm as long as you soak the roots well in a bucket of water before you plant them out.

In early autumn you can collect seed from perennials and hardy annuals and sow them outside. (Hardy plants are ones which can withstand several degrees of frost.) This will give them a longer growing period and a head start over similar plants sown in the spring, so they will flower earlier. You can also sow half-hardy annuals under cover. (Half-hardy plants cannot withstand very low temperatures, but they will survive a mild frost or two.)

From mid-autumn well into winter you can take hardwood cuttings to increase your stock of deciduous shrubs and climbers and some bush fruits, such as gooseberries and black, red and white currants. It is important to remember to take hardwood cuttings during the dormant season when the plant is not growing. The ideal time is just after the leaves have fallen in the autumn or during late winter to early spring, just before the new buds start to burst. Don't worry if nothing appears to be happening after you have planted your cutting; it usually takes quite some time for the cutting to develop roots and shoots and it may be well into the new growing season before anything starts to occur. For excellent advice on how to take hardwood cuttings look at the web site of the Royal Horticultural Society (www.rhs. org.uk).

Mulching can also be done now. The idea of mulching in the autumn is to trap the warmth and moisture in the soil. Therefore it is best done before the first frost arrives, but do make sure the ground is not dry - any rain falling after mulching will have another three or four inches to penetrate before it gets to the soil proper and will do little to sustain the plants. (For more on mulching see the spring section.)

DECEMBER TO FEBRUARY

By December the first frosts have usually arrived (even in the town or city) and we may even have had some snow. I don't really mind the snow simply because we don't usually get very much of it and when we do it doesn't lie around for too long. Some winters, however, turn out to be the exception that proves the rule, bringing snow and temperatures well below freezing for more days than I care to remember. When this happens, I can't help but think of the painting by Monet entitled 'The Magpie' which shows a wintry landscape, a hedgerow topped with a thick ridge of snow and a solitary magpie perched on a gate - a painting which can be found on many a Christmas card. I do wonder if there are any bee hives tucked behind the hedge, sheltered from the worst of the wind and snow. The garden doesn't seem to mind the snow too much, though. It acts almost as a blanket, protecting the soil and plants underneath it from dramatic changes in temperature. The bees aren't bothered too much about the snow either. They will be intent on keeping the queen alive and, having followed Craig's advice, you will have snuggled them down for the winter and done everything you can to help the queen and her entourage survive the colder months. Even so, you may see the odd bee flying on a warmer, bright day just to see what's going on and whether there is anything worth reporting back about. The critical temperature for individual bees is 6°C. If they venture out of the hive and the temperature falls below this level, they will die. There are plants which flower during the winter months but hopefully our bees will recognise that they are just tasters of things to come and not part of a proper menu.

Winter Plants for Bees

Winter flowering plants are few. Winter flowering plants that are suitable for bees are even fewer: the sweet-smelling viburnum (V.

bodnantense) is one such and mahonia (M. aquifolium) can be relied upon for a valuable pollen source during this season. The winter heather (Erica carnea) will also flower now as will gorse (Ulex sp). It's not really surprising that so few plants are flowering at this time of year - most insects are hibernating or in a state of near torpor which nature is well aware of, so there is not the same imperative to provide food.

Winter Jobs for Bee Keeping Gardeners

Gardeners, too, are often in a state of torpor, especially if they have imbibed too much over the festive period. It doesn't matter so much, however, because the number of jobs that can be done during the winter period is relatively small. As long as the ground is not frozen and is not too wet, bare-root trees and shrubs, especially roses (see the summer section for varieties) can be planted during this dormant period. You can also plant containerized plants at this time; they require the same conditions as bare-root material. In addition, continue to take hardwood cuttings (see autumn section).

Now is also a good time to take stock and see if any alterations can be made to an existing garden to encourage bees, or, if you are in the enviable position of planning a new garden, to decide what should be included to bring bees and other beneficial insects into your space. Look through this chapter and pick out some plants from each of the seasons to include in your garden, then consult the numerous seed catalogues that land on the door mat soon after Christmas. My plant-aholic desires have taken many a flight of fancy whilst poring over these seed and plant lists, only to be brought forcefully down to earth again when I tot up how many packets of seeds I have earmarked and, more to the point, how much they will cost. My long-suffering husband invariably points out, too, that our garden would have to be the size of a minor principality to accommodate all the plants that I have in mind. Ah well, that's the beauty of armchair gardening - the only limitation is the capacity of your imagination. So I shall doze off, seed catalogue in hand, tea at my side, and dream of balmy summer days when the sun is shining, the flowers are blooming and my bees are buzzing happily around me ....

MARCH TO MAY

Spring can creep up on you. It sometimes feels that whilst it was winter last week it is now spring. There is a Chinese proverb that says: 'Spring is sooner recognized by plants than by men.' It's true. Long dormant plants seem to identify before we do that the nights are getting shorter and the days warmer, and now is the time to start thrusting new shoots through the soil and letting buds burst from their casings. Bulbs start to colour the ground and scent the air; fruit trees are clothed with blossom far showier than any designer dress. And bees start to emerge from the hive - slowly at first, with scout bees being sent out to reconnoitre and report back if they find anything of worth.

Spring Plants for Bees

Early spring can be a difficult time, especially if we have had a severe winter. Precious food sources for the bees are reluctant to make an appearance. There should be little cause for worry, however. Although, unfortunately, some colonies do die over winter, if you have a strong, well-fed colony then it should survive until the new season brings a fresh source of food. Some of the first flowers to emerge in March and through to May are those which humans pass by without a second glance, but they are a beacon for bees. These are the flowers of the willow (Salix sp) which are an important source of pollen and can also provide modest amounts of nectar. Flowers which are hard to miss by both humans and bees are those of the crocus which do not give much in the way of nectar but are an excellent source of pollen. Other spring flowering bulbs and tubers will also provide a little food for bees, notably grape hyacinth (Muscari sp), the native bluebell (Hyacinthoides non-scriptus - our own British bluebell, not to be confused with the Spanish bluebell, H. hispanica) (opposite - top left) and the winter aconite (Eranthis hyemalis).

April can be tricky. Many early flowering plants have already had their day and if the weather is adverse then late spring flowers are unwilling to emerge. Mahonia aquifolium (opposite - bottom) can always

be relied upon to fill the pollen gap with vivid yellow flowers scenting the air. Before long, however, nature puts on one of her most attractive and alluring shows - blossom time. Apple (previous page - top right), pear and plum all come into flower in late spring and although orchards of top-fruit trees are certainly a sight to behold, you don't need to live in the country to see branches laden with blossom. Many an urban street is lined with cherry trees and even a modest back garden can be home to a dwarf fruit tree. Other trees are also flowering now but these are often overlooked as valuable sources of bee food. Horse chestnut (Aesculus

hippocastanum), hawthorn (Crataegus monogyna), holly (Ilex aquifolium) and sycamore (Acer pseudoplatanus), all trees to be found in an urban setting, provide pollen and varying amounts of nectar.

There are other plants whose flowers are available to bees during spring. Shrubs such as the ornamental quince (Chaenomeles sp), the shrubby, sweetly-scented honeysuckle (Lonicera fragrantissima) and early viburnum (Viburnum burkwoodii and V. x

judii in particular) provide pollen and nectar in varying degrees. Early flowering broom (Cytisus sp), the flowering currant (Ribes sanguineum) and rosemary (Rosmarinus) also help to break the bees' fast. Of all the shrubs in flower at this time of year, however, the gorse (Ulex sp) is the one that will be most attractive to bees. There is a saying that when gorse is in flower, kissing is in season - which is to say that it is always in flower! It is, therefore, perhaps one of the best bee plants, even if it doesn't come in the top five of every gardener's 'must have' shrubs.

You can find some perennials in flower in spring, the majority of which are quite modest like the bugle (Ajuga reptans), white dead nettle (Lamium album), forget-me-not (Myosotis) and pulmonaria. One useful perennial which is invariably grown as a biennial is the wallflower (Cheiranthus). It is a stalwart of spring bedding, providing underplanting for majestic tulips: it is also a valuable bee plant.

Spring Jobs for Bee Keeping Gardeners

Bees are not the only creatures up and about in spring time: that perennial of all specimens emerges, bleary-eyed from looking at those seed catalogues – the gardener. There is much to do in the garden in spring - digging, planting, sowing seeds, tidying up ....

Many gardeners do their tidying up in autumn, cutting back spent flowers, clearing away leaves and mulching. I tend to do this in the spring. I happen to like the somewhat disheveled look, mainly because the vegetation provides over-wintering sites for

♥

SOME BEE-FRIENDLY PERENNIAL PLANTS AND BULBS

♥   ♥

Agastache *(A. foeniculum)*
Autumn-flowering crocus *(Colchicum autumnale)*
Bee balm *(Monarda sp.)*
Black-eyed Susan *(Rudbeckia hirta)*
Bluebell *(Hyacinthoides non-scriptus)*
Bugle *(Ajuga reptans)*
Catmint *(Nepeta sp)*
Crocus
Dahlias *(single-flowered varieties)*
Fleabane *(Erigeron sp.)*
Forget-me-not *(Myosotis)*
Geraniums
Globe thistles *(Echinops ritro)*
Grape hyacinth *(Muscari sp.)*
Helenium
Honeywort *(Cerinthe major purpurescens)*
Jacob's ladder *(Polemonium caeruleum)*
Lamb's ears *(Stachys byzantina)*
Lavender *(Lavandula sp)*
Leopard's bane *(Doronicum sp.)*
Michaelmas daisies *(Aster novi-belgii and A. novae-angliae)*
Oriental poppies *(Papaver orientale)*
Perennial daisy *(Leucanthemum sp.)*
Pulmonaria
Sea holly *(Eryngium sp.)*
Valerian *(Centranthus ruber)*
Wallflower *(Cheiranthus)*
White dead nettle *(Lamium album)*
Winter aconite *(Eranthis hyemalis)*

insects, leaves will rot down into the ground anyway, and I don't always get round to mulching before the colder weather sets in. Tidying up in the spring disturbs fewer insects, and mulching at this time of year means that the winter moisture can be trapped in the ground just as it is warming up. I like to mulch a little bit early actually, when the spring bulbs are just beginning to show. They are quite happy to have another couple of inches of material to grow through and if it were left any later they would be far too advanced to be able to be mulched around. I can't make enough of my own compost from vegetable trimmings, dead flowers and the like but I am lucky in that I have a ready supply of well-rotted horse manure available from a local stable. I bribe Craig with as many cream teas as he can decently manage in one sitting to transport bags of the sweet-smelling stuff for me in his trailer. It really is sweet-smelling, honestly. You should be able to get hold of a handful, close your eyes and take a jolly good sniff without thinking of ... well, you-know-what. If you can't, then it hasn't rotted down enough. If you can't make your own compost or get hold of any manure then ordinary multi-purpose compost will do a reasonable job although it won't contain many nutrients (and no worms - vital for the well-being of your soil) and it will be a lot more expensive. If you do have to use it, then look out for special offers at garden centres and home improvement stores.

Spring is the ideal time to get digging and plant all manner of containerized hardy, woody and herbaceous plants. (Leave the annual bedding until much later when all risk of frost has passed. And certainly don't be tempted by special Easter offers at garden centres - even a late Easter is still too early to plant out tender annuals.) Shrubs are

♥

SOME BEE-FRIENDLY ANNUAL PLANTS

♥ ♥

Borage *(Borago officinalis)*
Californian poppy *(Eschscholzia californica)*
Candytuft *(Iberis sp)*
China aster *(Callistephus chinensis)*
Clarkia *(Clarkia sp)*
Cornflower *(Centaurea cyanus)*
Cosmos *(Cosmos bipinnatus)*
Heliotrope *(Heliotropium cultivars)*
Love-in-a-mist *(Nigella damascena)*
Mignonette *(Reseda odorata)*
Phacelia species
Poached egg plant *(Limnanthes douglasii)*
Sunflower *(Helianthus annuus)*
Sweet sultan *(Amberboa moschata)*
Zinnia *(Zinnia elegans)*

often overlooked, but they provide the 'backbone' of any border or planted area. In recent years there has been a move away from shrubs to ornamental grasses to provide such structure. Grasses do have an attractive form, and as the talented exponent of 'prairie' planting, Piet Oudolf, maintains, they do 'die' well, providing year-round visual interest, but from a bee's point of view they are worthless. Bees like flowering shrubs. Evergreen or deciduous, they don't mind, as long as they have pollen- and/or nectar-rich flowers.

Now is also the time to visit your local nursery. They will have top-quality stock which has weathered the winter or new season plants that are raring to go. Look for plants with strong new growth and with a root system that fills the pot, but is not pot-bound. Knock the plant out of its pot and have a good look - any nursery owners worth their salt won't mind you doing this.

(See Chart on page 119)

Some annuals (see chart above) and biennials (plants which in effect grow in the first year

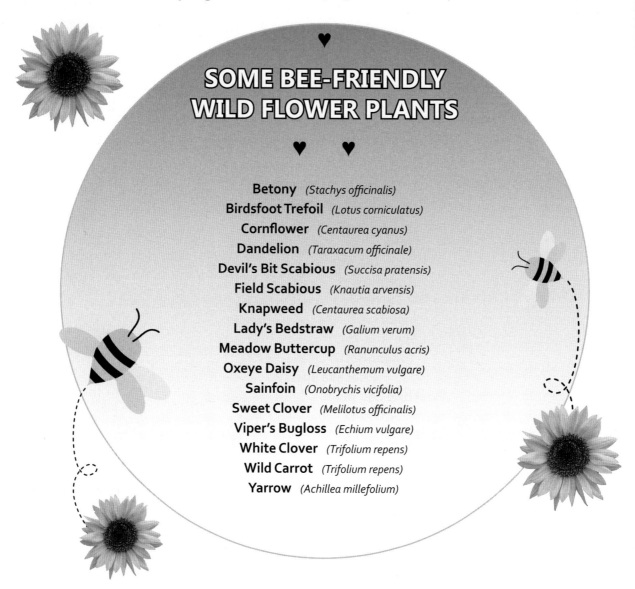

♥

SOME BEE-FRIENDLY WILD FLOWER PLANTS

♥   ♥

Betony *(Stachys officinalis)*
Birdsfoot Trefoil *(Lotus corniculatus)*
Cornflower *(Centaurea cyanus)*
Dandelion *(Taraxacum officinale)*
Devil's Bit Scabious *(Succisa pratensis)*
Field Scabious *(Knautia arvensis)*
Knapweed *(Centaurea scabiosa)*
Lady's Bedstraw *(Galium verum)*
Meadow Buttercup *(Ranunculus acris)*
Oxeye Daisy *(Leucanthemum vulgare)*
Sainfoin *(Onobrychis vicifolia)*
Sweet Clover *(Melilotus officinalis)*
Viper's Bugloss *(Echium vulgare)*
White Clover *(Trifolium repens)*
Wild Carrot *(Trifolium repens)*
Yarrow *(Achillea millefolium)*

and flower and die in the second) can be sown under cover now, although be sure to look at the cultivation notes on the packets of seeds to see which will benefit from such treatment. You don't need an elaborate greenhouse for this - you would be surprised how many small seed trays you can fit on a kitchen window sill!

Many hardy wild flower seeds can also be sown direct into the soil during late spring (a few suggestions are given in the box above). Even a couple of square feet given over to a 'wild' area in your garden will attract any number of beneficial insects as well as bees. I love wild flowers. An abiding memory I have is, as a child lying face up in a field of cowslips across the road from where I lived, one late spring afternoon. The honey scent and butter-yellow colour of those fairy bells will stay with me forever. Wild flower seeds are available in mixes or as individual species from a number of suppliers. (Look at www.crossmoorhoney.com for some suggestions.) But please, whatever you do, make sure the seeds you buy are from a reputable company which uses sustainable sources. We must not allow our plant heritage to be decimated by unscrupulous traders.

You can increase your shrub numbers by taking softwood cuttings during spring. Material is taken from the soft and flexible young shoot tips, which root readily at this time of year. As long as you can pot them by mid-summer they will develop enough roots to carry them through the winter, otherwise leave them and pot them up next spring. Again, for excellent advice on how to take cuttings look at the website of the Royal Horticultural Society (www.rhs.org.uk).

As the bees start foraging in earnest it is important to remember to provide a source of water for them in the garden. Bees will fly large distances to find water but it will be better for them if you can provide some close to the hive. A pond with a pebbled edge is a luxury which you don't need to aspire to, although it will attract no end of wildlife to your garden. Bees will naturally take water from wet surfaces such as pebbles, soil, grass and even wet washing, rather than from an open surface. They are not good swimmers so they need a good solid surface to land and walk on in order to access the water. In addition, in warm weather a strong colony will require up to 4 litres (7 pints) of water a day, so a good, clean, regular supply is essential. Any shallow, water-tight container filled with pebbles or crocks and topped up with clean water is ideal but I have found that a poultry feeder with some pebbles in the canal at the bottom where the chickens

would normally drink from is an excellent solution. Don't site it too close to where you sit or where you regularly walk, but put it far enough away from you and close enough to the hive to encourage the bees to use it as their watering hole.

JUNE TO AUGUST

If spring was the starter then summer is definitely the main course. Much depends on the weather, of course, but given even an average summer there is now a surfeit of food for bees and the nectar cup overfloweth! Bees will be out at first light and will continue collecting all day until dusk as long as the air temperature is satisfactory and rain keeps at bay. It's no wonder that during the main foraging season worker bees only live for about six weeks: they work every waking hour, flying on average 500 miles in their lifetime. The nearer the food source is to the hive, the less distance the workers have to fly during each foraging flight, the less energy is expended flying to and from the hive and the more pollen and nectar can be gathered. In an urban environment a huge variety of flowers is on their doorstep - or should it be hive-step? Now is the time when the majority of plants are flowering - ranging from trees through to shrubs, herbaceous perennials and fleeting annuals.

Summer Plants for Bees

Not many urban gardens are large enough to accommodate a tree, but trees are a mainstay of parks and amenity planting. During the summer months you will find sweet chestnut (Castanea sativa), the Indian bean tree (Catalpa bignoniodes), the tulip tree (Liriodendron tulipifera) and the lime (Tilia sp) amongst others, providing varying amounts of bee food. Lime trees were at one time the quintessential street tree but they have fallen out of favour because they can attract large numbers of aphids and scale insects which secrete 'honeydew', a sugar-rich sticky substance which falls on cars parked underneath the trees. Bees sometimes collect honeydew but they prefer to stick with pollen and nectar direct from the flower.

Many shrubs also flower during the summer season. Look out for abutilion (Abutilion vitifolium), cotoneaster, deutzia, lavender, mock orange (Philadelphus sp), snowberry (Symphoricarpos sp) and the guelder rose (Viburnum opulus). We must not forget roses proper, however - the nation's favourite flower. Gertrude Stein (the early twentieth-century novelist and poet) famously wrote

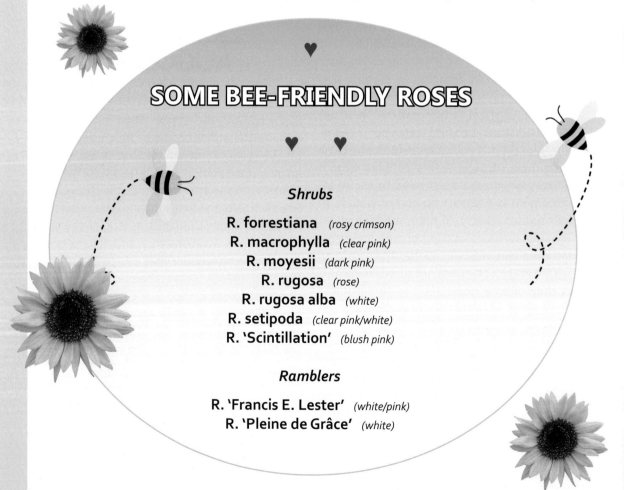

SOME BEE-FRIENDLY ROSES

Shrubs

R. forrestiana *(rosy crimson)*
R. macrophylla *(clear pink)*
R. moyesii *(dark pink)*
R. rugosa *(rose)*
R. rugosa alba *(white)*
R. setipoda *(clear pink/white)*
R. 'Scintillation' *(blush pink)*

Ramblers

R. 'Francis E. Lester' *(white/pink)*
R. 'Pleine de Grâce' *(white)*

in a non-horticultural context: 'A rose is a rose, is a rose.' But, horticulturally speaking, roses are not all the same. There are the big, blousy, old-fashioned blossoms; prim, hybrid teas; multi-bloomed floribundas and repeat-flowering English roses. These are undoubtedly beautiful but are of little use to bees. It is the original species roses and cultivated single-flowered types which bees need, and because the flowers are not sterile they will produce beautiful hips in the autumn. In addition to these suggestions there is also Rosa gallica officinalis, the Apothecary's rose, also known as the Rose of Lancaster. Although it's semi-double it does have bright yellow bee-attracting anthers and being a Lancastrian by 'adoption', I just had to include it.

(See Chart on page 124)

In early summer (and if the weather is clement, as early as May) you will see Leptospermum scoparium in bloom. Leptospermum, otherwise known as tea tree or manuka, is, for many people, the ultimate bee shrub. It is native to New Zealand where it can be found growing wild, but here in Blighty it is classed as half-hardy, requiring some protection during the winter months. Manuka honey is reputed to be one of the best medicinal honeys and can command a very high premium. However, unless you have a plantation of it (like the Tregothnan estate in Cornwall) then you can't guarantee that your bees have been feeding solely on that one sort of shrub so it is best not to get your hopes up about producing your very own manuka honey! Nevertheless it is worth growing one or two of these evergreen bushes simply because bees love them and they are good garden shrubs.

Best of all the native shrubs for bees during

Tips!

- *Ivy flowers in the autumn and is an extremely important source of nectar containing about 70% glucose.*

- *Bury your Daffodil bulbs at least 2/3 times their own depth.*

- *The gorse (Ulex sp) is a spring shrub and is the one that will be most attractive to bees.*

- *Leave the annual bedding until much later in spring when all risk of frost has passed.*

- *When purchasing plants, look for ones with strong new growth and with a root system that fills the pot but is not pot-bound. Knock the plant out of its pot and have a good look.*

- *You can increase your shrub numbers by taking softwood cuttings during spring.*

late summer and early autumn is heather. Commercial bee keepers will move their bees up to the heather moors during this period because they can guarantee that the heather will be the only flowers that the bees will feed on and can therefore sell it as pure heather honey. (Bees will fly up to three miles for food but if they can find it on the hive-step then they will not waste energy looking further afield.) There are in fact two shrubs that are known as heather: bell heather (Erica cinerea) and ling (Calluna vulgaris). It is the ling which flowers slightly later than bell heather which produces the 'real' heather honey. This honey is dark with an almost caramel flavour and is often described as the 'Rolls Royce' of honey. Obviously you won't want to turn your garden into a heather plantation but if you can include some then the bees will love you for it. A word of warning, however; both Erica and Calluna prefer acid soil which may pose a problem unless you grow some in a container where you can use specially formulated ericaceous compost for acid-loving plants.

Summer is when the perennials really take centre stage. A list of perennials beloved by bees would fill several pages, so I shall recommend just a few (in no particular order of preference) which I can guarantee will keep our buzzing friends happy. Agastache (A. foeniculum), catmint (Nepeta sp), many geraniums (not the tender summer geraniums which are really pelargoniums), sea holly (Eryngium sp) and fleabane (Erigeron sp) will all get the bees buzzing, as will globe thistles (Echinops ritro), heleniums, Leopard's bane (Doronicum sp), and the aptly-named bee balm (Monarda sp). Oriental poppies (Papaver orientale), Jacob's ladder (Polemonium caeruleum), lamb's ears (Stachys byzantina) and honeywort (Cerinthe major purpurescens) are also worth planting. Bear in mind, too, purple loosestrife (Lythrum salicaria), field scabious (Knautia arvensis) and oxeye daisy (Leucanthemum vulgare), all of which are wild flowers and therefore ideal for bees, but they also make beautiful garden plants.

Annuals too are at their most resplendent now and if you have a perennial patch with

a few gaps it is certainly worth filling them in with some of the annuals mentioned in the box above.

Herbs are mainly grown for their leaves but if you can let one or more plants flower then you will provide a welcome variation to the bees' menu. (And to your own, since a number of herb flowers, such as borage, chive, rocket, sweet cicely and lavender can be eaten and are a delicious addition to a summer salad.) I mentioned earlier that one of the best bee herbs is marjoram (Origanum vulgare), the flowers of which produce the most concentrated nectar known, containing 79% sugar. Hyssop (Hyssopus officinalis), mint (Mentha sp), sage (Salvia officinalis), summer and winter savory (Satureja hortensis and S. montana), lemon balm (Melissa officinalis

– the name is an indication of its worth as a bee plant, since melissa means honey bee in Greek), and lavender (Lavandula sp) all flower at this time of year.

Some plants which are grown as agricultural or horticultural crops and are sought after by commercial bee keepers because of their high nectar or pollen yield can also be useful in the domestic garden. I am thinking particularly of borage, also known as starflower (Borago officinalis) (see picture - page 125), sunflowers (Helianthus sp), lupins (Lupinus sp), clover (yes, I know it's thought of as a weed) and a whole range of fruit and vegetables. Borage is classed as a herb and is grown commercially mainly for its oil but the flowers are lovely in a salad and you can't have a proper Pimms in the summer without its leaves. Sunflowers are grown commercially for seed and oil and lupins are making a comeback as a fodder crop. Farmers will grow clover to increase the nitrogen levels in the soil but for the average gardener clover is an anathema. Forget your prejudices because white clover (Trifolium repens) is one of the best bee plants known. Red clover (Trifolium pretense) is beloved by bumble bees but is of less use to honey bees which have shorter tongues and cannot reach down the corolla tubes of the red clover flower to reach the nectaries. If red clover is cut once or twice, however, the subsequent flushes of flowers have shorter tubes which the honey bee can reach. Another plant whose common name is sweet clover or melilot (Melilotus sp) is of the same family as, but a different species from, white and red clover. It looks quite different from 'ordinary' clover in that it grows to a height of 30-50cm and has pea-like yellow or white flowers. It rates as highly as white clover in the bee plant chart.

Craig has stated elsewhere in this book that fruit and vegetable yields can be increased by anything up to 36% if bees are active on the crop. This is a huge advantage for farmers and growers. And if it is true for commercial crops then it is also true for domestic ones. Strawberries, gooseberries, raspberries, blackberries, currants - indeed all soft fruits - will all have an increased yield if pollination is high. Fruiting vegetable crops such as broad and runner beans will benefit (but not peas which are self-fertile and do not need external pollination), as will the cucurbit family of courgettes, marrows and pumpkins.

Summer Jobs for Bee Keeping Gardeners

Summer is the busiest time of year for the bee keeper. It is also one of the busiest times for the gardener but if you want to keep bees and tend your garden then you have to give priority to your buzzy friends – they are far more important. If you don't get around to doing those gardening jobs then it doesn't matter – there is no point getting upset. Gardening should be a pleasure, not a chore, and unless you totally neglect your plot for months on end then the worst that will happen is that the garden will begin to look a little unkempt. Never mind.

If you do have time, however, there are jobs to be getting on with. The risk of frost has long passed and the soil has warmed up enough to start planting out the more tender flowering annuals that you so assiduously sowed under cover earlier in the year and have been potting on since. Early summer is also the time to sow direct into the ground. Many annual seeds prefer this method but

once sown they dislike being disturbed. You must thin them out, of course, otherwise the growing space will become cramped and the plants will grow leggy and weak, but apart from this leave them alone and they will give you a beautiful display with a minimum of effort on your part.

If you have bigger gaps in your planting than a few annuals can fill, then pop along to your local nursery to stock up. The beauty of containerized plants is that you can plant them at any time of the year, except in the dead of winter if the ground is frozen or is too wet, so you can see them in flower at your local nursery before you buy. If you do plant anything during the summer, however, make sure that the ground is well prepared and that you water the plants diligently during the first season. Thereafter they should be able to fend for themselves. Talking of water – make sure there is still a sufficient source for your bees.

During late summer and into autumn you can increase your stock of all sorts of plants ranging from trees and shrubs to climbers and herbs (especially hyssop, lavender, rosemary, sage and thyme) by taking semi-ripe cuttings. These are cuttings from the current season's growth where the base of the cutting is hard while the tip is still soft. They will root easily at this time of year as long as the soil is not allowed to dry out. See the Royal Horticultural Society's web site (www.rhs.org.uk) for further information on taking softwood cuttings.

Weeding: as a gardener you either love or hate it. I would rather not have to do it but recognize that if I don't I shall be in trouble later in the season. I always think that I have better things to do with my time - like looking after my bees! - than hoeing off or digging up plants which happen to have chosen a place to grow that doesn't meet with my approval. After all, as has been

said many times before, a weed is just a plant growing in the wrong place. There is a way to mitigate the weed invasion, however, and that is to have mulched earlier in the year. (See spring jobs). Even if over-wintering weed seeds lurking underneath the mulch have germinated, there will be too much material for their skinny shoots to push through and they will perish before they reach the surface. Perennial weeds are a different matter. If they have a foot-hold (or should it be root-hold) in your garden then the only way to be rid of them is to dig them up, or if they are pernicious you may have to resort to a systemic herbicide. A systemic herbicide is a weed killer that you spray or paint on the leaves of the weed. The chemicals are absorbed by the plant and are taken down to the roots so the entire plant is killed off, not just the parts that are in direct contact with the killer. It does the job but you have to be careful that you don't accidentally get weed killer on plants close by that you don't want to get rid of.

As a bee keeper I am extremely conscious of using anything ending in '-cide', (from the Latin caedere - to kill). I don't know if one of my bees would be killed if I inadvertently sprayed it with a herbicide and I am not going to carry out an experiment to find out. I do know, however, that I would never use any insecticide or pesticide in my own garden. I care deeply about not only my own bees but all other beneficial insects, especially ladybirds, lacewings, hoverflies and other types of bees. (Wasps are perhaps the exception. Although they do have a role to play in pollination and aphid

control, some of them, on occasions, have been known to attack my bees and rob the hive of honey so they are definitely insects non-grata as far as I am concerned). If the ecology in your garden is balanced then there will be no need to use bug killers anyway.

And finally, there is something at this time of year which too many gardeners forget to do. Imagine you have weeded and watered, mown and mulched. Your bees are happily buzzing around collecting pollen and nectar and you have carried out all the 'bee' jobs that Craig has suggested you do during this season. What is there left to do? Nothing. Nothing? Just so. Take some time out. Sit in the garden. Look at the flowers. Listen to the bees. Just be. And enjoy!

CHAPTER NINE

A LITTLE BIT OF BEE HISTORY

.......back to Craig again!

If I hadn't been incredibly talented and handsome and the role of James Bond hadn't already been snapped up by Roger Moore then I probably would have taken on that role. As it was I had to look for summat (something) else in life. As you know I have led a varied life but there has always been a constant light in my life, and not just the bees. I have always been fascinated with history and how society's actions have impacted on mankind and the trail that has been left behind. If I had had my life all over again (apart from being a James Bond stand-in or even a stunt bum for Mel Gibson) I might well have been an archaeologist or a researcher of military history. You can look at small wars, industrial disputes or financial crises and yet despite crossing international boundaries, their impact is often only local. Even the extinction of a species may have little or no impact on the world-wide society. Let's look at the Dodo. Long hunted from the shores of Madagascar, its loss has not affected the UK or even Western Europe and, although a tasty bird, its demise isn't sadly missed. And I don't mean that in a blasé way. Yet the plight and history of the humble bee is something that has crossed continents, has a folklore all of its own and supported a huge industry. Bee history is not one that can be confined to a single place or time. Anything associated with

bees, especially honey bees, whether it's the financial fortunes associated with trade, disease, or wisdom, has transported itself into man's social history.

Man and the bee are probably the two creatures that have lived alongside each other uninterrupted since history began. Man has always valued the majority of the properties that the bee has had on offer: from the sting and its medical properties to the sweet tasting honey that was early man's first form of sweetener. Long before raw sugar cane was being brought over from the Caribbean, honey was traded for very considerable sums of money - it was mankind's only true sweetener. No rotting black teeth then. So can we pinpoint a time for the bee, for honey and for society's wrangling over the product? 12th April 1977. Only joking. What I can say is that the oldest bee 'in captivity' is to be found preserved in a blob of amber. The specimen is believed to be at least 80 million years old. As for man's involvement, we do know there are cave paintings in Spain believed to be around 6000 to 8000 years old which depict bees. It amazes me, purely from a bee keeping perspective, that man was recognising the importance of one creature and making crude paintings of it millennia ago. Man's fascination clearly has extended for thousands of years and for thousands of miles. Representations of bees were painted on the walls of caves in South Africa and in Lower Egypt in about 265BC. The latter representation depicted a bee keeper with 5000 hives – either that or the artist had a twitchy hand. In ancient Egypt honey and bees wax were staples in medicine and in the embalming process. As I mentioned earlier in the book the Greeks and Romans referred to bees as little messengers of the gods, but they were keeping bees and recording their findings 3000 years ago. At the first Olympics it was recorded that athletes used honey as a sugar boost just before a race and over longer distances they would take honey diluted in a drink to sustain them. Clearly the Greeks knew something we didn't. According to the Roman Pliny, drinking a glass of honey and cider vinegar every day cleans the system out and advances good health. It would certainly wake you up after a skinful of ale the night before.

What has religion had to say about honey and the friendly bee? Well, every book from the very early days of the Talmud right up until the Book of Mormon all mention the bee, honey and its properties and all mention it as a gift from God. But this reference to the bee and all things good isn't based purely around the children of the Abrahamic religions - it spreads further afield. Hindu gods, Buddhism and the Shinto gods of Japan all had opinions on the uses and beauty of honey and the bees. In Greek mythology the nymph Melissa would care for the baby Zeus. At this time Zeus was being hidden from his father. Melissa would plunder hives for their honey so that she could feed the infant Zeus. When Zeus's father, the king of the gods, discovered this Melissa was turned into an insect as an act of revenge. When Zeus had grown into a god in his own right he took pity on Melissa and turned her into a honey bee so that she could make honey for eternity. Amazing! If he was that grateful he could have turned her back into a nymph. Like honey the use of beeswax can be found throughout ancient history and mythology. Into the pre-Christian area we can find that wax was an offering to the gods as well as a sacrifice; it (beeswax) was used in the purification of the dead, circumcisions, marriage ceremonies, and births.

The ancients always had good strong moral tales, but after the Greeks and Romans disappeared, their tales remained and were translated into the society of the Dark Ages, yet very few new tales appeared to support the role of the bee. In the Dark Ages of Europe, monasteries kept their own bees, not just for honey to trade and consume, but because they used so many candles it was cheaper to keep their own bees for the wax. It was the monks who used fermented honey to make mead which is a wonderful

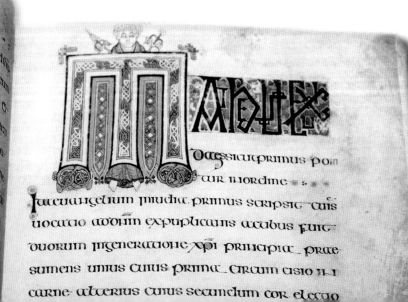

Traditional style European skep hive

explorers arrived was the absence of honey bees. All the pollination would have been done by wind, which was probably enough to pollinate the amount of crops needed by a small indigenous population, or by smaller insects chancing their arm by mindlessly hopping from plant to plant. When Erik the Red landed in Newfoundland he originally named it Vinland because of the number of grapes on the vine. In fact he was impressed by the total amount of produce available and presumably it wasn't just meat and fish, but also fruit, vegetables and other edible crops. And yet there were no bees to do the work of pollination. Bees were introduced from Europe to not just the Americas but also to New Zealand and Australia. By the mid 1600s records show that as the human population was growing along the east coast of America so was the bee population. As humans moved west, so did the bee so that, by the time Europeans reached California, the bee had been there for some time and was integral in making the West Coast of America the land of milk and honey - literally. It wasn't all beer and skittles for the native American Indians, however, as they referred to the bees as "white man's flies" or perhaps a more appropriate translation would have been "bugger me that hurts".

liquid concoction that retains the beautiful flavour and sweetness of honey but with the sharpness of the alcohol. A great way to while away those long winter nights - a beaker of mead to imbibe and the latest blockbuster parchment to lose yourself in. Honey and beeswax have been used for trade for centuries and the monasteries were at the heart of this trade. When you think about it, it was reasonable for them to do so since it was the monks that were the go-betweens between factions; it was the monks that had the land and the knowledge; and it was the monks that often controlled the trade routes. They also needed wax and honey and propolis to use in writing. Many of Europe's great books, for example the Book of Kells, were completed by monks. And as a resource beeswax and honey were integral to the production of these works of art. One of the oddest things noticed in both North and South America when the early

We know that for thousands of years, if not longer, man has been interested in, if not totally fascinated by, the workings of the bee. I don't know if it's a little bit of fear that attracts people to the hive; running the gauntlet in a way, wondering will I get stung or won't I? Perhaps if bees were as docile as kittens people would not be interested in them at all. (It's a little like crime, I think; take away the fascination and illegality of it and in many ways it stops or at least loses some of its allure.) The history of bees and their interaction with mankind hasn't

better insulating properties. I don't know.

A certain Mr Hitler had offered my father the unique chance to travel the world free of charge, to visit strange and wonderful lands, to kill people he had never met before and catch weird and unpronounceable social diseases. My father was with the British Army for many years in Palestine and across north Africa (we always referred to it as 'lost in the desert'). He loved the countries he found himself in and had a great respect for the Bedu and spent many months being taught about water supplies, irrigation and water sourcing, hence his knowledge of the bees and the water. Always one to ask questions, he was in Palestine and was talking to a farmer who kept bees, at the same time enjoying a long, typical Arab offer of hospitality: mint tea, various lamb dishes and, of course, the obligatory hubble-bubble pipe. On careful examination of the farm what took my father by surprise was the material used in bee hive construction. Mud. Mud, glorious mud. The Arab chappie was shaping his mud into large earthenware jars and as they were unfired, they were easily constructed and dried quickly. They were then placed on a wall that had sections removed on a grid - a bit like Bob Monkhouse's Celebrity Squares, but dozens and dozens of them. Once the pot had dried in the sun it was placed in the grid and marked. Bees were placed inside and the process started all over again, year after year.

gone unnoticed. But it hasn't always been about trade or conquest - the history of bee keeping has a fascinating vernacular history in the way bees have been nurtured, kept and transported. If we go back thousands of years to a time when mankind was becoming proficient in bee keeping, he was using a hollowed-out tree in which to house the colony. Well that's almost what a wild colony would use. But if you had a good year and the bees wanted to make more honey than the tree stump could accommodate, you were struggling. In parts of the Baltic States it is traditional to use a carved piece of tree trunk as a bee hive. This may be because in this part of the world the bee season is short and the natural bark and timber fibre have

My father passed his love of travel and foreign climes on to us all. (Unfortunately his passion didn't extend to my brother who became a member of the flat earth society, believing the world did not extend beyond St Helens in the south and Southport in the north. My sister embraced my father's love of travel and adventure and set forth on independent and exciting journeys. She got as far as Germany where she met her first husband and then married him in Mobile, Mississippi. She is now on German husband number three and engagement ring number four.) As for me, I travelled as much as I could and as far as my meagre wage would take me. What I do know is that whilst on my travels around the world I was passing through the Bekka or Beqqa valley in The Lebanon. What a fantastic and fascinating country, with really polite and humble people who only ever want to talk to you and befriend you. On my visit to the valley I remembered my late father's tales of Arabs and hospitality. I journeyed out of Beirut and up country to where the final point of the Great Rift Valley ended at the Bekka valley. I had left the war torn, battered and bruised city behind and ventured forth. I can truly say that, despite moving forwards it felt as if I were going backwards. In time, at least. Yes, there were cars and the obligatory Honda 50cc with three people sitting on it. But the further I went, the more I seemed to journey into the past, and the more I enjoyed it. I eventually stopped and washed my face and feet and

The Bekka Valley, Lebanon

saw some children playing. I thought at least one of them will speak a few words of English. I smiled, gave them some toffees and said a few words of English. They all said "hello". When I asked them where I could sleep, they said "hello"; then I asked them if they would give me a million dollars and they said – yes, you've guessed it. So I went off to find an adult. The children followed me, but as the night was drawing in and it was still and warm I didn't think it was too much of a problem if I ended up sleeping under the stars. I was taken into a shop that sold food. A single light bulb hung from the ceiling and the flies rolled around the light making the place look the more like a Moroccan brothel than a grocery shop. I asked in my best English whether he sold honey; he shrugged his shoulders. I then tried the same sentence in German and French but neither worked, so I made the sound of a bee buzzing around and then made the gesture of spreading something with a knife on a pretend piece of bread, and then making the sound of pure delight. "Ahhh!" said the shop keeper and, grabbing my hand he took me outside across the square to a friend's shop. I made the same noises and acted out the same scene with the knife and imaginary bread. The second man smiled in acknowledgement,

turned around and passed me a can of fly killer. He charged me well over the odds and shut the shop. I wondered about my father's tales of legendary Arab hospitality and if it had been damaged by 500lb bombs dropped on Iraq by Uncle Sam and good old Mother England.

As I sat in the shade of a cypress tree and considered where to stay, I wondered at this renowned hospitality. I looked at myself in a small travel mirror and realised why I might have been treated thus. I was dusty, looked as bad as I smelt and my sartorial elegance was totally absent. But I loved me, if no one else did. I brushed myself down, changed my t-shirt and had a quick wash (more of a splash, really) with some bottled water. I got a book out on bee keeping and started to read when a young man of a similar age approached. His English was workable and he obviously wanted to practice. We spoke of the war in Lebanon, the Falklands war, the problems with Syria and I realised that his life had been brought up in pure conflict, where mine had had a minor blip. He pointed to a picture of the bees on the front cover and laughed. Making buzzing noises and flying around he called me over to see something. Behind a wall about 200 yards away there was a grid and in each space was an earthenware jar of bees, lying on its side - there must have been 200 of these. You could hear the gentle hum of the colonies at rest on the warm night. I was taken to see the old chap, the grandfather of the boy I had met, to tell him of our connection and how his method of keeping bees had been described to me by my father all those years ago. It was then and

there that I realised that the Arab hospitality that my father had encountered was alive and well and living at the top end of the Rift Valley in a small village east of Beirut. I was supplied with dates and lamb and all sorts of incredibly sweet desserts and cakes. Almost all of the desserts were flavoured with honey from the old man's bees.

After dinner we continued to sit on the floor and nod and make noises that appeared to communicate a dislike of this and a love of that, all the time his grandson translating as best he could. My grandfather wants your opinion on his honey he said and offered me several to try and look at, to smell and sniff and continue with the 'ooohs' and 'aaahs' of pleasantries. I dipped my fingers in one that tasted of caramel and one that tasted of butterscotch. Then I stuck my finger in one which was dark, almost black like molasses. I remarked that it was a beauty amongst honeys – I had never tasted anything like it. My host glared at me and started to babble away to his grandson ten to the dozen. I smiled like an idiot hoping they were saying something complimentary about their guest. In fact they were very annoyed because I had just tasted a honey made from palm trees and the sticky nectar that comes before the dates arrive. The grandson explained it was called zadar, and this was the cheapest honey they could produce and was mainly given to their livestock when they were young or ill. Only this morning it had been fed to several of their young camels because they were sick, I was told. The fact that I had sampled it without being offered it and enjoyed it more than the grandfather's other honeys was sacrilege. I couldn't apologise enough.

In the morning the grandfather invited me to see some other hives. Naturally I offered to give him a hand so we set off in a battered old Toyota. Eventually we stopped for tea and the old man asked, through his grandson, whether I knew how to use a gun. I explained that I had often used a shot gun shooting clays (this posed a bit of a problem for his grandson to translate, but we got there in the end, I think). He actually wanted to know whether I could use an AK47 properly. I said I had never used one but would be sure that I could get the hang of it. I thought we were off for a little bit of east/west male bonding, when in fact this chap had bee hives in a valley over the border in Syria. Now I've never been to Syria before, let alone without a proper visa or even a passport - and especially carrying an AK47. I could see the headlines in the Liverpool Echo: "Local man declared stupid by Foreign Office". It's official. I explained my predicament as best I could to the pair and they dropped me off - sans machine gun. I wandered back to the village, shooting things en route, but this time only with a camera.

CHAPTER TEN

ΓOLKLOΓE . MYTHS . LEGENDS AND SOME ΓACTS

"Three Cheers for St Ambrose"

"Hip Hip Buzzzzz"
"Hip Hip Buzzzzz"
"Hip Hip Buzzzzz"

Over many years of being a bee keeper I have changed my mind about a number of things, but one constant that has run through my life is that I have been very fortunate. Here I am again looking to the bees for inspiration. Sitting next to a hive with a nice cup of tea, the breeze in my hair and a couple of bob in my pocket there is nowt wrong with the world. How had I come to this fortuitous position? I am a great believer in the turn of a coin, the movement of an arm or the flick of a switch. That movement will take you one way or the other. You look one way and you miss seeing the woman of your dreams, you look the other and you see your future unfold before you.

As I sat there in a field with the rapeseed, a carpet of warm yellow sunshine, and the pine trees in their glory swaying in the breeze, the smell of pine drifting and tantalising my nostrils, I realised that if I had not become a chef and savoured the joy and appreciation of food and its creation I may never have become a bee keeper. If I had never become a lawyer I would never have become a bee keeper and, most extraordinarily, had I not had cancer I most certainly would not have become a bee keeper. I would have been at the same old 9-5 every week, every year. (Although it never was 9-5, it was always 6am until midnight.) And now here I was, sitting with my new found friends - friends who wanted to sting me if I bothered them too much. Alone with 75,000 friends, a cup of tea and my future in front of me.

As a lawyer I was always a pragmatist. Never one for superstition. If there wasn't fact or evidence to enforce a conviction then it didn't truly exist. I had heard all the tales

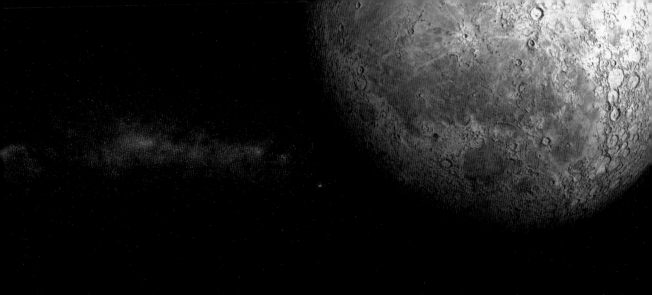

of bee keeping: the history, and even the history of the history; that it's the world's second oldest profession; the singing to the bees; the death of the bee keeper. Blow that, I thought, stick with fact. Over the years, as I have become more in tune with the bees, I have started to delve into the world of the myths, legends and ultimately the folklore of the world of bee keeping. I have to be honest and say that it has fascinated me, this new world of superstition, magic and respect. I realised that it didn't matter which culture, which religion or society, old or modern you looked at, bees and honey played a major part. There is a town not far from where I live called Blackburn. Over the years I have passed the gates to Corporation Park many times. A proud and beautiful set of sandstone pillars support the park's gates and over the top of the gates, just where the key stone sits, is a crest and on this crest sit two bees. I wondered how many locals knew of this. In fact the several I asked all knew what it meant; they even told me you could see the bees on the crest of the town's football shirts. On further examination I discovered that the bees were there because of the town's connection with the cotton industry and the product coming from the Americas. Blackburn was known fondly as "Cottonopolis". It drew its wealth from the cotton imports and the garments made in the many, many mills in and around the town: just as the bee is seen as the epitome of skill, perseverance and industry, so too were the workers. It's not just honey and fruit and vegetables that we need to thank the bees for, but also the garments we wear. Because without the cross-pollination of the cotton bolls we wouldn't have the benefit of cotton underwear and I would be sitting here rather uncomfortably.

Over the years I would often laugh at folklore until I was approached by an elderly farmer. He asked me how my last two years had been. Now I am used to people asking me about bees - the general public is fascinated and I am happy to answer their questions. But I wondered why this exact question, why did he ask about the last two years? I thought maybe he had read something, wanted to talk about CCD or tell me about a friend who kept bees in the 1940s. My curiosity got the better of me - why specifically the last two years? Well, he said, you won't have any success with bees when there are thirteen lunar months in a year. I thought about this but was still none the wiser. He went on to tell me that being so close to the coast the draw of the moon on the tides brings in too much salt air and moisture to the coastal plains. As long as you have bees in this area you will be scuppered boy, he said. I asked him if next year would be the same and should I give it all up. He said that two consecutive thirteen lunar month years won't happen again for another 75 years. They are very rare, he said. I breathed a sigh of relief: by that time I shall be long gone and someone will have already told my bees about my demise. This year promises to have a better summer - and there are only twelve lunar months - so we shall see.

Well this is local folklore but it must have come from somewhere and it's the somewhere that interests me. Whether you look at the Romans, Carthaginian's, early Christians or Islamists, they all identified with the healing properties of honey and the mystique of this creature that produces such a sweet product yet has the capability of inflicting such pain and at the same time brings about its own untimely death. If you look back towards the ancient Greeks or Romans, their early traditions concluded that bees were the messengers or little servants of the gods - they were there to do the bidding of those on high. The Romans believed that a swarm of bees would bring bad luck because, as they were on the move and as they were the servants of the gods, the latter were quite clearly upset about something and had ordered the bees to enact their displeasure. So the sight of a swarm was thought to be an omen of doom when in reality

it's quite the opposite (unless they settle in your hair or in your trousers, of course). I quite like the story that says that Bahus was the world's first bee keeper. He apparently domesticated them whilst on his travels to Frakia. And in the legend of Jupiter he was said to have been fed and protected by the bees when his mother Rea hid him in a grotto on the mountain of Ida. You have to remember that the Romans and the Greeks didn't just sit down and invent a load of traditions; invariably they borrowed a good number of them from other, earlier societies. Quite often these were adapted from the likes of the Draconians and the cultures of the Middle East. Ancient Egyptian culture too played a major part in the development of Roman and Greek society and customs. In ancient Egypt the sun god Ra was credited with creating bees from his own tears. I like this one especially - to think you could be so vain as to believe that something so pretty and industrious could be created thus. In addition to this the Egyptians would embalm their dead in honey, but only the nobs, not the common folk. And at the side of the

Egyptian hieroglyphics

corpse they would leave an offering of honey to help them on their journey to the afterlife. As well as honey, mead or honey wine was a drink always associated with the afterlife or at least staving off its affects. An ancient drink, if not the oldest one, it was believed that its properties could promote immortality - but again, it was only available to the upper echelons of the society. Are you seeing a picture develop here? Only gods - or those who thought they were - had direct contact with bees and any of their by-products.

The ancient Greeks believed that bees would always be seen at a blacksmith's forge because they liked the sound of banging metal and were therefore associated with the love of music. (I think it is more likely they were after the heat from the forge.) I decided to set up my own experiment to test the bees' preference, if any, for music in its widest sense. I set up my CD player alongside a hive and set to work. I first played part of an early Joni Mitchell. (Aaah Joni, when will you marry me?) I then played a classical piece, Albinoni's Adagio in G Minor, and then the Sex Pistols: all for ten minutes and at the same volume. With Ms Mitchell the bees were active and interested in the sound waves coming off the CD player; they were OK with the quieter, gentle songs. The music of Albinoni was also accepted, except when a high note or series of high notes were played. And as for Mr Rotten the bees were less responsive; they completely ignored the Sex Pistols which I thought would have had the opposite response. I believe it was because bees are annoyed and disturbed by the sharp peaks in sound reverberation, whereas something at a lower ratio and consistency seemed to appeal to them. Although not a scientific experiment it would seem that the bees were less upset by lower, deeper notes than the higher shrill

ones. I have also come to the conclusion that Johnny Rotten was born of virgin birth.

I like the folklore that says that bees are associated with, and symbolic of, sexuality, chastity (amazing that you often find them juxtaposed like that), purity, almost of a virginal quality, care and fertility. Yes that's me all over. I could impregnate a chair by just sitting on it. Early Christians believed that bees were capable of virgin births and therefore became symbolic of the Blessed Virgin and interwove their presence into early Christian history, practice and belief. This is a good example of the idea that if you don't understand how something works or functions, it's easy enough to associate it with religious belief to 'explain' how it has come about.

All cultures appear to worship to some extent the role that bees and honey play in their society or cultural norms. In India the Hindu gods Vishnu, Krishna and Indra were often referred to as the nectar-born ones (Madhava). In ancient Egypt, along with Indian and Greek cultures, the bee is seen as the divine spirit of the soul. In other cultures it is seen in different lights. In ancient Celtic culture, for example, bees were seen as bringers of great wisdom and as being messengers from one world to another. The Minoans believed them to be symbolic of rebirth, immortality and regeneration. Closer to home, though, in the United Kingdom, regional folklore abounds. (I have to say that a large proportion of Cornish folk believe in an independent Cornwall and that the county of Kernow should be recognised, so for those of you reading the book in Cornwall I apologise for lumping you together with the rest of us.) In Cornwall it is folklore that a bee keeper must advise his bees that he is moving them in advance so they can ready themselves. It is also believed that if a bee keeper does not inform his bees of a move they are entitled to sting him harshly and, should he move them on Good Friday, the stings may result in his death. I wonder if the bees would inform his wife of his death. Again in Cornwall your bees would be free to leave you if you should speak to them harshly and, as messengers of god, should you swear at them they will leave. I'm surprised my bees have not all disappeared years ago - but then I'm not a Cornishman.

Wales appears to be abundant with tales of all sorts; dragons, elves and especially death and bees. In Wales a flight or swarm of bees would indicate mixed omens. But I have no doubt that the sight of a swarm would be a terrifying observation for any uninformed bystander especially after all the films and books about "killer bees". The thought of all those stings and an unsettled swarm that might at a moment's notice dive on you and render you the colour of spam is truly terrifying. A swarm that gathered in a dead tree was a portent of a death in the family. And yet an early swarm might mean an early crop of honey which appears to look towards the positive. In Wales you should never sell a hive on a whim, but a hive given away as a gift will bring the new owner good fortune

and a good supply of honey. When a colony decides not to leave the hive you are not to worry because it indicates forthcoming rain. (This is for me the most unfeasible of all folklore because if you knew Wales as well as I do, then bees would never be able to exist there, let alone make any honey. It is a beautiful country with wonderful people and friendliness unbounded. But damp.) Apparently a virgin can walk safely through a swarm of bees in Wales. I would love to know how they came to that conclusion. I have many times had to deal with swarms and I can assure you my days of wearing white have long since past and I have never been harangued by bees. Another Welsh bee-inspired piece of folklore is that if you work with your bees at night time they will not sting you. Again, untrue. There are numerous other sayings, for example: a hive that is lethargic indicates a poor relationship with the owner and future misfortune; if you find a bee buzzing around your house, then the bee is a forward notice that a visitor is coming shortly; if a bee flies over a child having a kip then the child will have a blessed and fortunate life. The latter saying isn't just confined to Wales – it is commonplace in most of Europe as well as the Steppes and the Eastern Anatolian region of Turkey. I have no doubt that this tale has come from further afield, perhaps along the Silk Road?

Myths and legends about honey bees and the stories that are attached to them are widespread throughout the world and it is because of this that a possible book and television series is being looked at where I follow the footsteps of the ancient Silk Road and travel from Istanbul to Xi'an in China and examine the process of transporting honey from one continent to another along with the spices and silks. Along this route through the old Soviet republics I

will examine how apiculture and honey have influenced the countries of the region and how some of the myths, legends and folklore have remained local, whilst some have passed along the Silk Road and into western civilisation. In Turkmenistan where it was believed that bees would only live in a harmonious family, the concept of telling the bees the news is widespread. As most families in Turkmenistan had a bee hive of their own in order to trade honey on the silk road or to use as a sweetener for their food, it was important for the family to be able to maintain their hive and so one of the myths that grew up was that the bees should be told everything that happened, both good and bad. The bees were considered to be models of domestic harmony and peace and to be highly industrious workers whose attributes were highly sought after, so the family that kept its bees alive and had a good, heavy honey crop were able to advance the idea that they (the family themselves) were a mirror of the bees, and not the other way around. Bees that thrived only did so because the family was good. This piece of folklore was advanced even further in Turkmenistan when bees were to be told of a family death or the consequences would be that they too would die. In addition, any bad news had to be given before the sunrise of the following day if all was to be well with the hive.

I just wonder whether this is a form of covert social control - having a clean, happy hive was something to strive for and be proud of, which in turn reflected the state of the household and by extension village life itself. The folklore appears to be extended from not just telling the bees of a death, wishing them well and perhaps remarking on what the bee keeper's wife might be up to today, but to offering them gifts, leaving offerings outside the hive, almost turning them into deities,

becoming a cult of the bee. And that worries me as we never know where to stop.

In English medieval society when there was a death the bringer of the sad news would say a little rhyme to the bees: "Little brownies, little brownies, your master/mistress (insert name here) is dead". This had to be said three times to the hive and the idea behind this is that the bees, through their supposed association with God, would grant longevity and good health to the bereaved family. The custom of telling the bees has been carried through the years. I have always been told that when a bee keeper dies his bees need to be sung to and told of the death. You would communicate with these deaf creatures through the mystery of song (they 'hear', or rather feel the vibration of the sound, rather than the sound itself). You don't have to sing any particular song, any will do, and then inform them of the death. The bees would then come out at night and fly around the hive as a last act of respect for their former master. Sounds lovely. This made me smile as I thought what would happen if you sang "Smells like Teen Spirit" by Nirvana or "Back in Black" by AC/DC. Would my little friends be upset by the music or appreciate the kind thought that they had been informed of their master's death? I suspect that if I sang to them they would probably commit suicide.

What I can say is that over the years my opinions have changed. I no longer look at the role of the bee and the impact folklore has had on bee keeping with such legal, clinical disdain. I have not gone as far as leaving gifts for my 'employees', but I have started talking to them more often. But bees don't like vibration, so talking to them can upset them. When we are working the bees we are either silent or speak in hushed tones. I have told them of the deaths in the

family that mattered. Although my bees never met my father or brother in law it was them that I told when they had passed away and I don't really know why, it just seemed the right thing to do, which I suppose reflects the change in my outlook towards the bees and their social history. When my father and brother in law died it came as a great shock to the family. My brother in law was a bobby for thirty years and my father a soldier at El Alamein and Monte Cassino: they were both brave, solid upstanding men. Yet when I went to visit them in the Chapel of Rest they lay alone in their new suits and new coffins and they had this great journey to make and I couldn't help or assist either of them in any way. I placed a small bee in each of their coffins. Such a symbol of strength, of ingenuity and industry; solid upstanding members of the insect world. I knew that as I entrusted their souls into the arms of Jesus that this little bee would guide them and protect them.

Once the bees have been told of a death they must be invited to the funeral, be given a piece of funeral cake after the funeral, and perhaps a glass of wine should be left next to the hive; not to do so would insult the bees - they will follow the coffin and then turn around at the door in disgust for not being formally invited and a poor honey harvest would befall. Well, I did invite the bees and offer them cake, but they didn't come. I didn't invite them into the car and it's a long bus ride to Burnley. In fact I'm sure they wouldn't be welcome on the 152 from Preston to Burnley. But hey, what a way to get a seat all to yourself. "No, I'm sorry you can't sit there, that's the bees' seat" - who is going to argue? Although I have not gone as far as telling them of births and marriages or news from over the seas, I always say hello and goodbye to my hives, but most of all I thank them for their contribution and I thank them for their honey. It costs nothing to be kind.

It is interesting to note that as cultures looked more and more towards a patriarchal form of society and religion, bees and the role of the queen were lost and replaced by the role of the king bee. In fact, until the early 1830s the Queen bee was referred to as the King bee because it was thought that only a male could rule such a complicated and well-ordered kingdom. The implication was that a female couldn't possibly control a hive. (Tickles me pink, that one.) Yet a plethora of societies and cultures have acknowledged and worshipped earth mothers and have seen a parallel with them in the role of the queen bee in the colony. Over countless millennia we (society) have taken the bee and its actions to represent that of teamwork and collective responsibility and perseverance. It is because of those collective attributes that the basic tenets of many 'matriarchal' religions are centred round the worship of an earth mother or collective form of womankind. The bee is an ancient symbol of sacredness and is associated with the earth goddess, mother earth or the divine feminine. This belief is associated with the queen bee's responsibility for the hive and colony which is representative of the enclosed, nurturing space of the womb. Goddesses in ancient societies, Greek and Roman alike, have all been associated with the bee and

the hive; names that will be familiar to you have all been referred to as the queen bee - Diana, Persephone, Rhea and Aphrodite, to name a few. Aphrodite was worshipped at a honeycomb-shaped altar at the base of Mount Eryx. The altar was symbolic of the holy geometric shape of cosmic order and harmony. The honeycomb represented the perfect union of the macrocosm within the perfect microcosm.

As my wife is forever telling me, she is the quee bee and it's my role to worship at her feet. However, I am at liberty not to tell her of the honeycombed altar. She will have me making one in the back yard if I do.

Honey was thought to have magical properties. I suppose a substance that has to be collected in the way it does, running the gauntlet with the bees with the chance of being stung just for something to sweeten your tea with can take on a miraculous aura. Honey was seen as such a magical substance that once fermented and turned into mead it was considered to have prophetic powers which gave it value far beyond its original worth. In fact all the

the bee, but it's interesting to note that the bee features famously in both Oriental and Islamic culture and is noted in many Aboriginal beliefs. (I feel a second book coming on!) There are some very interesting and hard facts, however, although we generally know very little about our chums. One story I can tell you (but can't mention any names) had to do with an aircraft manufacturer. The person I knew who worked at this particular establishment told me that his company had spent a lot of money and time on research into looking at bees, especially bumble bees. Now you might question why an aircraft manufacturer would want to have an insight into the workings of a bumble bee? They wanted to develop a helicopter that would fly quietly and into a headwind with no problems like a bumble bee, but could take a huge payload with a minimum amount of wing span and a minimum amount of fuel. So, the massive aircraft corporations are looking at the workings of bees to help them develop future aircraft, perhaps even with a military use. Bunkum, you might say. Well, the American military are spending a substantial amount of money trying to train bees to detect land mines – seriously. I don't know if that's a good or a bad thing.

properties and by-products associated with the hive have been imbued with sacred and mystical properties: royal jelly was believed to be only for the lips of royal children and bee pollen and beeswax were used in many religious and magical ceremonies in ancient cultures.

Bees were extremely important to the development of western societies right up until the seventeenth century. Honey was regularly used as a sweetener for foods and drinks and was used in the fermentation process for cider and ales, but when cane sugar came along it knocked honey off its pedestal and as the importance of honey declined, so did the folklore and the telling of the magical tales that surrounded it. Recently, however, there has been a resurgence in the use of honey and with it the myths, legends and folklore are all being revived.

I've really only taken a Western approach towards the myths, legends and folklore of

Let's look at some of the facts associated with both honey and bumble bees, beginning with honey bees first. When you take a teaspoon of honey in your mouth, think of how many honey bees it has taken to get it to this stage. Well, a honey bee will spend its entire life working to produce a twelfth of a teaspoon. So, for a hive to produce the pound of honey that you find in the supermarkets, it involves thousands of bees all working as a collective. Bees must produce at least 60 pounds in weight of honey for the hive to survive over the winter. Which is why I put two gallons of inverted syrup (artificial bee food) on the bees as soon as the heather honey has been taken off in early September, otherwise my buzzy chums would starve.

A honey bee can make up to 2000 visits to flowers looking for pollen and nectar a day. Obviously she can't carry that much nectar and pollen from that number of flowers in one trip, but she will visit many flowers in one sortie before returning to the hive. A single round trip can be anything up to six miles, and if you are working in kilometres and the metric system, then it's still six miles. Now flying that far and being that small obviously puts a huge strain on the body and can cause irreparable damage. So a forager bee may only live for three weeks. Sometimes you may see a bee crawling along the ground. This can be for a number of reasons: the bee may simply be exhausted, in which case you can give it some sugar to get it on its way; or it may be close to death, especially if it's completed the amount of work we just mentioned. It may appear kind to stand on a bee that is slowly dying but I don't feel this is fair. I am wondering now that if I do kill a bee, do I have to go back to its hive and tell the rest of the colony.

How fast does a honey bee fly? Well, about 15 mph. You might think with all that pollen and nectar that's fast; it is if you are trying to swat one, but in the insect world it's rather slow. Honey bees are made for short trips, not long distances - a bit like a short-haul 737 compared to a long-haul airbus. Their wings

must flap 12,000 times per minute just to stay upright and moving.

A queen honey bee will mate and hold within her a life-time's sperm. It takes her just 48 hours to be mated and a lifetime to lay. That's not bad for a small creature that may lay 1500-1600 eggs a day for several months and will live for up to four years. Admittedly she won't be laying as many eggs in her third year, and even less in her fourth year, but she will still be laying. So we're looking at a total figure of about a million bees born in her lifetime. She can lay her own body weight in a day and because she is a laying machine she has no time for herself; the worker bees do the rest of the labour for her. The worker bees within the colony have specific jobs to carry out. There are attendant bees that bathe, groom and feed the queen; nurse bees who care for the young bees and foragers who go out on a daily run looking for pollen and nectar. Certain bees guard the front of the hive (landing board) and other bees are busy building up the foundation of wax. Housekeeping bees clear the cells and arrange the order of things whilst the undertaking bees remove the dead and the deposited chalk brood. All this goes on just to keep one queen in the manner she was shackled to. What of those boys, I hear you shout. Drones are the equivalent of the couch potato. They are born to do one thing; fertilise the queen and once they have, they die. What a price to pay.

Unlike humans bees use all of their brains. A bee's brain is packed with a million neurons and its brain is measured in cubic form. I know what it is in metric but not imperial, so we will say the bee's brain is a cubic 'very small thing' (about a cubic millimetre). This 'very small thing' is capable of directing the bee to do a huge range of things. One of the most

interesting things, I find, is their 'language' skills. By this I don't mean that they can speak a dozen different bee languages or dialects, but they do communicate with one another in the most extraordinary way: they 'dance'. This dance is known as the waggle dance and it was not until 1973 that its code was discovered and broken. The bees perform this dance to tell other bees about successful foraging sites, watering sites, pollen and nectar sources and even about new housing locations. (Yes, said one of my bees the other day; I would like to move upmarket and live somewhere like Whalley, near the old Monastery - there's a lovely four bedroom detached with its own pool. Such abilities these creatures have.) For a while it was thought that bees had two distinct dances, one where they moved in circles to indicate sites close to the hive, and the other where they waggled to show where distant sites were located. After cracking the code we now realise that the circular dance is in fact a short waggle dance with a short waggle run. So what is a waggle dance and can I have one for Christmas. The dance consists of from one to one hundred or more circuits, depending on how long it takes the worker to convince her sisters of her discovery. Each circuit consists of two phases, the waggle phase and the return phase. The worker that carries out the dance will initially start off by running through a

small figure of eight pattern (the waggle phase). This is followed by a turn to the right to circle back to the starting point (the return phase), where she collects £200. Another waggle run follows, followed by a turn to the left to complete the circle, and so on. The duration and direction of each circuit and the actual position of the 'dancing' bee are extremely important because they contain the directions and distance from the hive to whatever the worker has found. We also know that the 'dancing' bee produces two alkanes (tricosane and pentacosane) and two alkenes, which they release into the air

Let's say a bee has found a good source of water which is required by the colony. If the source is directly in line with the sun, this is represented by waggle runs in an upward direction on the vertical combs. Any angle, whether to the right or left of the sun, is coded by a corresponding angle to the right or the left of the upward course. I like how the next piece works: the noted distance from the colony to the source is identified in the length of time the waggle dance takes. If the source is further away, then the longer the waggle run is. For those mathematicians out there, the exact calculation is a rate of seventy five milliseconds per hundred yards. What happens if a bee returns from a source and it is later in the day and the sun is not far off setting? The bees that are absorbing the information from the waggle dance will adjust the angles of flight to take account of the changes in angle of the sun. So, for example, the information about a source of water that a bee comes in with at 9pm on day one in the summer will be translated and recalculated for a first flight at 4am the following day. Amazing. My father spent many years in the Sahara and he told people he could smell water and could find it merely by the sniff of his nose. In reality he would

watch for bees. If there were no bees he would tell people there wasn't any water close by. If he saw a bee he would follow it to try and locate a water source such as a wadi, oasis or mountain water course. He told me of a time he watched some bees heading for a small but deep pool not far from where the wadi had been. The bees kept hovering and then backing off. What had happened was that the pool had been deliberately contaminated with crude oil. The bees knew it and gave the warning. I am pleased to say that my father made an improvised water feeder for the bees with some stones and a billy can.

An old lady come to the farm once and asked if she could place her hand in a hive as she was convinced that a bee sting would help her arthritis. Several stings later she walked away and thanked me. Now the idea that a bee sting will help with arthritis is a moot point. I know people who swear by the effects and people who just swear. But I wonder how many times and how often you need to be stung for it to have an effect. These days you can purchase a plethora of honey products or capsules with bee venom in them which seems to me a less drastic course of action. I recently read in an article that an American university hospital in St Louis was doing research into using bee venom in treating cancerous tumours. What a fantastic breakthrough if that dreadful disease could be halted or at least controlled by our buzzy friends. I have also read about the use of bee venom in the treatment of neuralgia, high blood pressure and even MS. There is, in fact, a treatment for MS that involves the use of snake venom so I wonder if there is a correlation?

You might come across the word "propolis". I mentioned it earlier in the book and bee keepers generally refer to it as bee glue. As bees generally don't like draughts or exposed areas that may allow vibration or unwanted visitors in, they will bung up a hole or area with this reddish coloured material that will harden and provide excellent support material. It is in reality a sticky tree resin that is mixed with wax to make it pliable. You can actually eat it and I have come across a large number of Muslims that like to chew it for health reasons. It is readily available from health shops. The other substance that as a non bee keeper you hear a lot about is Royal Jelly. Now I think the title is rather inviting and comforting. Royal Jelly is a milky white substance composed of digested pollen, honey and nectar all mixed together with a chemical from the nursery bee's head. It's the ginseng of the bee world. I said earlier that this substance is fed to all bees for a three day period, but after that it is only the queens who get further rations. It is incredible that this feeding regime can turn an average every-day worker into a queen. It's beginning to sound a bit like a West End musical already.

And finally. What is honey? Honey is basically bees' vomit. I try not to say this too often to too many folks because it puts people, especially children, off and lowers my sales considerably!

THANKS AND ACKNOWLEDGEMENTS

I really wanted to take this opportunity to say thank you to a number of people.

To Mole for all her hard work, unstinting support and cups of tea, not to mention her brilliant editorship and the chapter on the bee-friendly garden. You are a good friend and an excellent bee keeper.

To The Good Life Press for their encouragement, vision, dedication and terrible, terrible tea.

To Rachel Gledhill for her hard work, her professionalism on this book and her clear vision on layouts and direction. But 0 out of 10 for forgetting to make tea, Rachel.

To Phillip Thomas, my oldest friend, terror of the nation and excellent tea and breakfast maker.

To Mr Roger Tongue, my old school master who, in 1975, introduced me to bees as a 12 year old and allowed me to fall in love with them.

To my father, Uncle Nipper, Peter Du Rose, Uncle Steam Train, Joan Robinson and Barry Robinson, who never really had the opportunity to discover the happiness I now find with the bee.

To The Rosemere Cancer Trust at the Royal Preston Hospital, Lancashire, England for saving my life and the countless thousands thereafter. Keep up the good work.

To the farmers who have valued the use of bees on their land and have encouraged bee keepers nationwide.

To Joni Mitchell who has been with me since childhood and has yet to make me a cup of tea.

Craig Hughes - picked a fight with a bee
.........and lost!

INDEX

PHOTO CREDITS

- Disease images on pages 43, 44, 45, 53, 54, 56, 57, 58, 60 - **Crown Copyright - Food and Environment Research Agency, National Bee Unit.**
- Images on pages 48, 49, 69 - GNU Free Documentation License - wikipedia.org.
- Image on page 80 - GNU Free Documentation License - wikimedia.org.
- Supersedure cell pic page 64 - Copyright Kimberley Burch/Sunset Publishing Corporation.
- Image page 81 - Copyright - Rachel Gledhill.
- Image of Bekka Valley, page 137 - Copyright - Brian McMorrow.

All other images courtesy of the author or fotolia.com.

All efforts have been made to credit copyright. If there is anybody we have inadvertantly omitted, please contact the publishers and we will amend the credits in the next edition.

The Good Life Press Ltd.
The Old Pigsties
Clifton Fields
Lytham Road
Preston PR4 0XG
01772 633444

The Good Life Press Ltd. publishes a wide range of titles for the smallholder, 'goodlifer' and farmer. We also publish **Home Farmer,** the monthly magazine for anyone who wants to grab a slice of the good life - whether they live in the country or the city. Other titles of interest include:

A Guide to Traditional Pig Keeping by Carol Harris
An Introduction to Keeping Cattle by Peter King
An Introduction to Keeping Sheep by J. Upton/D. Soden
Any Fool Can Be a.....Dairy Farmer by James Robertson
Any Fool Can Be a.....Pig Farmer by James Robertson
Any Fool Can Be a.....Middle Aged Downshifter by Mike Woolnough
Build It! by Joe Jacobs
Build It!....With Pallets by Joe Jacobs
Craft Cider Making by Andrew Lea
Flowerpot Farming by Jayne Neville
How to Butcher Livestock and Game by Paul Peacock
Making Country Wines, Ales and Cordials by Brian Tucker
Making Jams and Preserves by Diana Sutton
No Time to Grow? by Tim Wootton
Precycle! by Paul Peacock
Raising Chickens for Eggs and Meat by Mike Woolnough
Raising Goats by Felicity Stockwell
Showing Sheep by Sue Kendrick
The Bread and Butter Book by Diana Sutton
The Cheese Making Book by Paul Peacock
The Frugal Life by Piper Terrett
The Pocket Guide to Wild Food by Paul Peacock
The Polytunnel Companion by Jayne Neville
The Sausage Book by Paul Peacock
The Secret Life of Cows by Rosamund Young
The Sheep Book for Smallholders by Tim Tyne
The Smallholder's Guide to Animal Ailments Ed. Russell Lyon BVM&S MRCVS
The Smoking and Curing Book by Paul Peacock
Worms and Wormeries by Mike Woolnough

www.goodlifepress.co.uk
www.homefarmer.co.uk